THE GENOMIC
REVOLUTION

THE GENOMIC
REVOLUTION

UNVEILING
THE UNITY OF LIFE

Michael Yudell and Robert DeSalle, *Editors*

JOSEPH HENRY PRESS
WASHINGTON, DC

with the
AMERICAN MUSEUM OF NATURAL HISTORY

Joseph Henry Press • 2101 Constitution Avenue, N.W. • Washington, D.C. 20418

The Joseph Henry Press, an imprint of the National Academy Press, was created with the goal of making books on science, technology, and health more widely available to professionals and the public. Joseph Henry was one of the founders of the National Academy of Sciences and a leader in early American science.

Any opinions, findings, conclusions, or recommendations expressed in this volume are those of the author and do not necessarily reflect the views of the National Academy of Sciences or its affiliated institutions.

Library of Congress Cataloging-in-Publication Data

The genomic revolution : unveiling the unity of life / Michael Yudell and Robert DeSalle, editors.
    p. ; cm.
Includes bibliographical references and index.
 ISBN 0-309-07436-3 (alk. paper)
 1. Genetics—Popular works. 2. Human genome—Popular works.
 [DNLM: 1. Human Genome Project. 2. Genome, Human. 3. Genetics, Biochemical—methods. QH 447 G33608 2002] I. Yudell, Michael. II. DeSalle, Rob.
 QH437 .G46 2002
 611'.01816—dc21

2002004016

Printed in the United States of America.

Contents

Part I
Genome Science and the New Frontier

Part II
Applications of Genomics to Medicine and Agriculture

Part III
Exploring Human Variation: Understanding Identity in the Genomic Era

Part IV
Financial, Legal, and Ethical Issues and the New Genomics

Foreword

In September 2000 the American Museum of Natural History was proud to host a landmark conference, "Sequencing the Human Genome: New Frontiers in Science and Technology." For two days we gathered an unparalleled group of experts, including Nobel laureates, distinguished moderators, and leaders in the scientific and business worlds, to focus on one of the most revolutionary and complex scientific developments in history—the completion of the first draft sequence of the human genome.

This milestone brings with it enormously compelling opportunities to better understand human health, our origins, and our relationship to other living things. At the same time, it raises profound ethical questions about issues already known and some not yet even imagined that will affect each and every one of us in such areas as the cloning of human beings and other species, the development of new medical treatments, privacy, and the criminal justice system.

As we begin to explore the Age of the Genome, there is a pressing need for public discourse on this vitally important topic. It simply cannot be for experts only. The American Museum of Natural History is uniquely positioned to begin extending these dialogues outside of the laboratories and scientific community, bringing them directly into the classrooms and living rooms of our country and the world. By doing so, we aim to share not only scientific understanding with the public but, equally vital, awareness of social implications and enhancement of the public's capacity to make both ethical and policy judgments.

Throughout its more than 130-year history, the Museum has occupied a critical place at the nexus of scientific research and public education, making scientific discoveries and interpreting them to the public. Never has this role been more important than on this topic, genomics, at this time, the dawn of a new century—the century of biology.

The American Museum of Natural History has long been a leader in developing new scientific fields and intellectual pursuits. Modern anthropology was born here under the leadership of Frans Boas and Margaret Mead. Paleontology found a new, more vigorous voice here, and, most recently, the Museum unveiled a new scientific and educational initiative of cosmic scale with the opening of the Rose Center for Earth and Space in February 2000.

We move now from the vastness of the outer reaches of the universe to the microcosmic inner workings of earth's organisms, living and fossilized. Today the Museum stands poised to take a leadership role in the crucial area of nonhuman genomics—crucial because the human genome itself cannot be fully understood in isolation.

The Museum's leadership role is especially important because our own genetic stuff simply does not tell the entire story of life on earth. The human genome alone does not reveal the relationships among species, human and nonhuman, the diversity of species, or the evolution and organization of life. The fossil record and growing frozen tissue collections housed at the Museum, including genetic information from both extant and extinct species, are essential to understanding where we were in the beginning so that we can appreciate where we are today.

Nonhuman genomics carries enormous implications for advancing our understanding of the behavior of individual genes across species, including humans, as well as for such urgent concerns as conservation and medicine, providing a road map that when used correctly will provide unbounded opportunity for better stewardship of our planet and all its inhabitants.

The field of genomics is uniquely suited to the strengths of the American Museum of Natural History with its collection of over 30 million specimens, one of the largest in the world, which forms an unparalleled record of life on earth. The Museum is home to over 200 research scientists, who, like their predecessors, gather and interpret evidence of the earth's history and evolution and the phylogeny of species. Our facilities and resources include state-of-the-art molecular laboratories, powerful cutting-edge parallel computing capacity, and a new frozen tissue collection, with capacity to house 1 million tissue samples.

We aim now to use the Museum's collections in a wholly new way to create a fuller, more comprehensive picture of the tree of life. With this research agenda as a foundation, the Museum is also undertaking an innovative program to educate the public about genomics. This is consistent with our mission—over the years the Museum has tackled subjects of enormous public interest and concern, including infectious disease, global warming, and species endangerment.

The September 2000 conference marked the beginning of a unique, sustained, and integrated effort to highlight and explicate the field of genomics. A full year of activities followed the conference, including a conference on parallel computing and a symposium on conservation genetics. Particularly important to its educational role, the Museum opened a groundbreaking special exhibition on genomics in May 2001. Entitled "The Genomic Revolution," the exhibit offered a comprehensive look at the science and issues of genomics from conservation and privacy to future prospects for the human race. It also offered a primer to the public and established a foundation on which to build deeper understanding in the years ahead. "The Genomic Revolution" will travel to venues throughout the United States, with a possible international tour to follow.

At the time of the opening of this exhibition, the Museum also launched a new Institute for Comparative Genomics. The Institute is a pre-eminent center for collections, research, and training in the field of non-human comparative genomics and pursues seminal research in the study of gene variation. This work informs our understanding of the human genome, the evolution and history of life, and the conservation of Earth's biodiversity.

It is not entirely clear to anyone where this genomic revolution will lead. But it is obvious that each of us has an enormous stake in understanding and managing the implications of this new era of scientific discovery. We are honored to have had the opportunity to include so many leaders of this scientific revolution in the two-day conference and in this publication, which presents their remarks.

Ellen V. Futter
President, American Museum of Natural History

Preface

 This book began as a discussion between us over five years ago, just as genomics was becoming an integral part of molecular biology and as our work evolved into genomic terrain from our respective disciplines of evolutionary biology and public health. At that time we had hoped to develop a conference on the genome at the American Museum of Natural History (AMNH) and eventually publish its proceedings. Always on the cutting edge of scientific research, museum exhibition, and scientific programming, the AMNH was, we believed, uniquely positioned to host a symposium on the scientific and public impact of genomics. The ongoing debates about the consequences of the genome, coupled with rapidly advancing genetic technologies, suggested to us the importance of increased public awareness of these issues. We drew up a wish list of the most accomplished names in a wide range of fields related to either the scientific or social aspects of this burgeoning science.

The symposium was designed, following the mission of the AMNH, to act as a nexus between the scientific community and the public and to translate what are complex and often inaccessible ideas to a common parlance. By hosting such an event, we had hoped that the Museum could become a model for the popularization of genomic research and also be a participant in what has become the most significant scientific undertaking of our time. To our great delight nearly everyone we invited to speak to the Museum public at "Sequencing the Human Genome" said yes. What a wonderful two days they were in September of 2000, listening to the provocative and thoughtful comments of the distinguished group of speakers whose words now grace the following pages.

The essays that follow are intended for both a lay and professional audience, and all do a great job of exploring the many aspects of genomics in a way that should not intimidate science-phobic readers. Some essays are more technical than others. Yet while the information in this book is challenging, don't let this deter you—it is presented in straightforward fashion. The book is divided into four parts plus our "Introduction," which is a look at the development of the AMNH's exhibition on genomics. Each part of the book is introduced by a science journalist who shares his thoughts on the state of the genome and where he thinks this technology is taking us.

Long in the making, this book is born of the diligent efforts of many colleagues, associates, and friends. And there are many to thank. This project would not have been possible without our contributors, all of whom rewrote their delivered addresses for inclusion in this collection. We generously thank them all. Stephen Mautner, publisher and editor of the Joseph Henry Press also deserves special thanks in bringing this book to publication. Maron Waxman, Special Publications Director at the American Museum of Natural History, shepherded this book from start to finish, and we owe her our ongoing gratitude for her interest in our work and her friendship. Kathi Hanna's painstaking work with us on the editing of the text brought the book together in its current form, and we owe her thanks for helping craft the richness and clarity of this volume. Finally, we'd like to thank C. Namwali Serpell, editorial assistant at the Joseph Henry Press, who guided us through the completion of this text.

We would also like to thank Compaq for its generous support in making the "Sequencing the Human Genome" conference possible. Here at the Museum many who helped realize the conference deserve special thanks. American Museum of Natural History President Ellen Futter's vision for the Museum as a home to both cutting-edge science and innovative and informative public programming means that "Sequencing the Human Genome" was only a part of the Museum's effort to educate the public about genomics. That effort continues with the Museum's exhibit on genomics, also shown at other museums, and the growth of genomic work in its laboratories. These types of efforts continue unabated thanks to President Futter and Museum Senior Vice-President and Provost of Science Michael Novacek. They both deserve special thanks for making all of this possible. We also owe our deep gratitude to Museum Vice-President Lisa Gugenheim and Elizabeth Werby in the Government Relations office who helped to make the conference a reality. Finally, special thanks go to the following Museum departments for their work on the conference: Development, Communications, the National Center for Science Literacy, Education, and Technology, Audio/Visual, Central Reservations and Ticketing, Custodial Services, Security and Safety, and Facilities Operations.

We would also like to thank David Rosner, all of the members of the DeSalle Molecular Systematics Lab at the AMNH, and the faculty and students in the History of Public Health and Medicine Program at Columbia University's Mailman School of Public Health for their helpful suggestions in the making of this volume.

Michael Yudell
Rob DeSalle
New York City
May 2002

THE GENOMIC
REVOLUTION

Michael Yudell
Rob DeSalle

Making the Genome Public

The American Museum of Natural History and the Coming Age of Genomics

 The genomic revolution has arrived. The results of the Human Genome Project—the 3.2 billion base pair long sequence of nucleic acids—are unveiling the fundamental elements of human biology. In the twenty-first century, genomic innovations will invariably bring about radical changes in medicine, agriculture, and the study of our evolutionary heritage.

The public has been captivated by the seemingly endless possibilities of genomics, so much so that "double helix" has quickly entered our common parlance. Yet most Americans remain remarkably unfamiliar with the realities of the genome. For example, a recent national Harris Poll indicated that only 50 percent of Americans could correctly identify that "DNA is what genes are made up of." Even fewer could explain its significance.

In the fall of 2000, as part of its ongoing mission to bring cutting-edge science to the public, the American Museum of Natural History held a two-day conference to examine the social and scientific implications of the human genome. "Sequencing the Human Genome: New Frontiers in Science and Technology" was the first major public forum to examine these implications following release of the draft sequence of the human genome.

Renowned scientists, including two Nobel laureates, bioethicists, historians, biotechnology entrepreneurs, and others participated in a mix of lectures and panel discussions. These presentations explored the ramifications of the Human Genome Project and addressed the social, economic, and ethical impacts of advancing genetic technologies and their effect on our understanding of natural history. This volume represents the fruits of that effort.

The conference was only a first step for the museum as it entered the genomic age. In the Museum's molecular laboratories, scientists are now integrating genomic technologies into studies of evolution and natural history. In May 2001 the museum unveiled a major temporary exhibition designed to present this revolutionary field to the public. This essay will explore some of the important themes of that exhibition and discuss how the often intricate and abstract scientific language of genomics was translated into a comprehensive exhibit for the public.

The American Museum of Natural History has a distinguished and long-standing tradition of making science and scientific discoveries accessible to the general public. For well over a century, the museum's halls, replete with fossils, models, and dioramas, have been home to a diversity of exhibitions that, with few exceptions, have centered on objects—exactly the fossils and dioramas that fill the museum's galleries. These object-driven exhibits utilize the charisma of a specimen to engage the visitor. An ancient Barosaurus specimen standing on its hind legs and towering 40 feet in the air does just that in the main rotunda of the museum every day. Once this visual connection to the specimens is made, the conceptual aspects of an exhibit can be presented. In the case of the Barosaurus, the museum can discuss a wide range of such dinosaur-related topics as predation, evolution, and extinction. The specimen draws in the visitor, but precisely because of that charismatic attraction he or she leaves with a much deeper understanding of dinosaurs.

"The Genomic Revolution" exhibit approaches the art of exhibition making and museum education in a much different fashion. Instead of relying on the allure of an object, the genomic revolution itself, in its entire abstract and complicated splendor, is what will attract the visitor. Here the physical specimens are secondary to theories, ideas, and scientific premises. The challenges for the exhibition team therefore were in translating these difficult concepts into dynamic and decipherable objects that illustrate the genome. To meet this task, a team of museum scientists and exhibition specialists, as well as a distinguished multidisciplinary advisory board, grappled with the problems for well over a year before delivering the final exhibit.

The exhibit advisory board considered several key concepts as necessary components of "The Genomic Revolution." These included, most prominently, that the visitor comprehend (1) the enormity of our genetic material; (2) the fact that despite this enormity, all humans are 99.9 percent genetically identical, and that through common ancestry we share an astonishing number of genes with all living things on earth; (3) the fact that our genetic code is an extraordinarily complex part of what makes us human and that this complexity interacts in very subtle and dynamic ways with our changing environment; and (4) that advances in genomics will be followed by considerable medical breakthroughs as well as significant social challenges. It was left to the exhibition team to integrate these abstractions into tangible objects.

The first task—to illustrate the sheer magnitude of our genetic material—was probably the easiest from an exhibitor's point of view. Still, this was a potentially difficult concept for a museum visitor to grasp and an important one too. Despite our rapidly advancing technological and theoretical insights, the immensity of our genome suggests that it will take time for genomics to produce results. Most people are surprised to discover that the unraveled complement of their DNA from a single cell extends 6 feet end to end. Moreover, the nucleic acids of our DNA, represented in letters printed in a 12-point font would literally *stretch* from Penn Station in New York City to Union Station in Los Angeles. To convey this scale, upon entering the museum's gallery the visitor sees three large plasma screens with DNA sequences, as they would appear on an automated DNA sequencer. It would take 11 months of continuous staring at the screen for a visitor to see all 3.2 billion base pairs contained in his or her genome. To help drive this point home, the visitor is also met by a stack of 142 bulky phone books filled cover to cover with Gs, As, Ts, and Cs—these 142 volumes containing 3.2 billion letters.

We are a physically diverse species. People literally come in a wide variety of shapes, sizes, and hues. Yet from a genomic viewpoint there is little intraspecies diversity. On average only 0.1 percent of our DNA varies from individual to individual. Our genes tell a very different story about human differences from our traditional understanding of human races. This point has not been lost on both natural and social scientists who wish to eliminate a biologically driven understanding of racial difference. Craig Venter and Francis Collins have publicly noted that our genomes illustrate that so-called racial differences are not discernible at the genomic level.[1] At Celera, for example, scientists were unable to differentiate between the genomes of individuals who had self-identified as Caucasian, African American, Asian, or Hispanic. The reason, according to Venter, is that "on an individual basis you cannot make that determination. You can find population characteristics, but race does not exist at an individual level or in the genetic code."[2]

Scientists have not always thought this way. For example, early in the twentieth century many scientists supported the eugenics movement. Eugenics—the belief that certain negative and deviant social behaviors are hereditary and genetic and can be correlated with particular ethnic and racial populations—encouraged "the socially disadvantaged to breed less—or, better yet, not at all."[3] The American Museum of Natural History was one of the world's most prominent institutions involved in the eugenics movement. For a time during the 1910s and 1920s the museum openly advocated and supported eugenics, even hosting the Second International Congress on Eugenics in 1921.[4] That congress, which included an exhibit on eugenics, attracted many of the world's most distinguished scientists and played an important role in popularizing and propagating eugenic theories and practices. In his opening address to the congress, Henry Fairfield Osborne, president of the museum, noted paleontologist, and prominent booster of early eugenics, said:

> To know the worst as well as the best in heredity; to preserve and to select the best—these are the most essential forces in the future evolution of human society.[5]

The effects of eugenics were far reaching and had an impact far beyond the narrow confines of academic circles where eugenics was widely celebrated. In the 1920s, for example, U.S. federal immigration restrictions were supported by eugenicist sentiment. Widespread sterilization laws across the United States also were inspired by eugenic sentiment. Between 1900 and 1935 approximately 30,000 so-called feebleminded Americans were sterilized "in the name of eugenics."[6]

"The Genomic Revolution" exhibit is an example of just how far both the museum and society have traveled since the days of the eugenics movement. Many leading genome scientists are paying careful attention to the history of eugenics. Referring to eugenics and its importance in the modern scientific consciousness, Craig Venter has said that "it is easy to look back on science and see the foolishness. It is very difficult to look forward and see it."[7] However, while most scientists have rejected eugenics and accepted that race is not a biological fact, some scientists and the general public hold fast to traditional racial ideology. It was the hope of the exhibition team that, by acknowledging the museum's role in the eugenics movement and by contrasting that role with the science of genomics, visitors would have a context in which to begin to understand current thinking on this often sensitive subject.

The exhibit's stated position—that the only race is the human race and that there is no biological basis for race—is presented in the section of the exhibit titled "99.9%," so named to highlight the amount of DNA that any two unrelated humans share. To illustrate this point, the exhibition team used some old-fashioned mendelian genetics with a genomics twist. We know that all humans are 99.9 percent similar and therefore 0.1 percent different. That means that between any two individuals plucked randomly from anywhere on Earth there will be approximately 3 million base pair differences. We know from mendelian principles that a mother and her biological child will have on average 1.5 million differences. These differences increase between generations and relatedness. For example, a grandmother and her grandchild will have 2.25 million differences while biological first cousins will have 2.625 million differences. What is so interesting and important to note is that our two randomly chosen individuals will have nearly the same genetic relatedness as biological second cousins, who have approximately 2.906 million base pair differences. In some sense this shows the visitor that all humans are family and that our perceived differences are basically meaningless at the level of the whole genome. This erosion of genetic similarity with familial distance illustrates the nature of the human family tree and helps visitors comprehend the nature of genetic differences among humans.

As part of this overall message on genetic sameness, the exhibit also explains the genetic relationships among all species, both living and extinct. The "Evolutionary Continuity Wall" surprises visitors by showing them, despite incredible physical differences, just how much we have in common with the world's other 1.7 million named species. We look nothing like mice, yet we share 90 percent of our genes with them. Other surprises include the zebrafish, with which we share 85 percent of our genetic material; the rat, 90 percent; and the fruit fly, 36 percent. Even roundworms share 21 percent of their genes with humans, and *E. coli*, bacteria found in our digestive system that are essential for survival, share 7 percent. Together the "Evolutionary Continuity Wall" and "99.9%" lead to a deeper understanding of the natural world. Practical medical advances coming out of the Human Genome Project will have the greatest impact on people's lives. But we should not forget that evolution plays an essential role in this process. Comparative genomics, an emerging field that identifies genes and gene function by comparing closely related species, marshals evolution for the genomic cause.

Genomics is also becoming a tool for both evolutionary and conservation biology. Several stations in the exhibit exemplify the importance of this type of research and the ways in which it touches the natural world. In a section of the exhibit called "The Profusion of Life," dioramas of different species tell a diverse set of stories. Scientists are studying herring gulls to determine whether exposure to oil spills is inducing mutations that are passed on to offspring. Raccoons and striped skunks tell the story of scientists using DNA to track down distinct strains of the rabies virus. And DNA evidence reveals that Florida manatees, now nearly extinct, have low levels of genetic diversity, which increases their sensitivity to disease and climate change. In another section of the exhibit, called "DNA Detectives," the use of genome technology in the fight against the illegal wildlife trade is highlighted. In an effort to reduce the pressures on wildlife, officials turn to DNA analysis to identify products made from endangered species.

The most important component of the exhibit, conveying to visitors the complexity of our genes—what genes do, how they do it, how they interact with our environments, and how genes make up our genomes—was probably the hardest to visually develop and is the most challenging to the public. Because people think of genes in such discrete and reductionist terms, the exhibition team had to construct a series of interactive exhibits that could facilitate a change (or advancement) in the museum public's understanding of genetics. This challenge comprises the majority of the exhibit and takes the public on a fascinating journey through the genome.

The journey begins with some basic science exploring the role that genes play in color vision. "How Genes Work" features a tour through the human eye all the way to the molecular level. An animation depicting models of an eye, a cone cell, the X chromosome, the opsin gene, the DNA sequence of that gene, and the opsin protein shows how cone cells must function correctly for a person to see in red and green colors. The animation also shows how mistakes at the molecular level in these cells, in DNA, can cause color blindness.

Errors in the arrangement of opsin genes in our genomes cause color blindness in approximately 10.5 million Americans.

The microarray station, the visitor's next step into the science of genomics, allows one to experience a revolutionary genomic technology while illustrating how genes contribute to disease in humans. It is also the exhibit's centerpiece as it connects the science of the genome to the future medical applications of genomics. The exhibit uses the story of breast cancer genes to help the visitor better understand the genetics of this dreadful disease. Through this example the microarray station explains how the genetic architecture of breast cancer will be used to improve our understanding of the natural history of this disease and to develop better treatments (both preventive and therapeutic). The microarray in the exhibition is an 800× magnification of the surface of a standard microarray chip. In practice a microarray allows scientists to analyze the activity of thousands of genes simultaneously. The chip modeled in the exhibit actually holds 8,102 genes and allowed researchers to compare healthy breast cells to cancerous ones. The microarray allows researchers to find genes that are active *only* in the cancer cells. This new tool will eventually help scientists discover news ways to diagnose and treat breast cancer. Microarray technology is not limited to this particular disease and will be an effective tool in studying and developing treatments for cancers and other diseases. Some scientists, for example, predict that one day microarray technology will allow doctors to develop therapeutics tailored to an individual's genome.

The myriad social implications of genomics are an integral component of the exhibit and allow the visitor to continue exploration of the complexities of the genome. To engage visitors in the possibilities of genomic medicine and science, several exhibit stations confront them with choices that they or other people in future (and in some cases present) situations might have to make regarding genetic technologies. The ethical questions presented are not meant to be exhaustive but rather to lead people through the problems inherent in these advancing technologies. The section on genetic testing nicely illustrates this approach, exploring several social issues, including privacy, uses of genetic knowledge, and prenatal testing.

These subjects are used to examine some of the complexities of genetics, such as the ways in which genetic abnormalities are detected, the likelihood of and ways in which these abnormalities are passed from generation to generation, and the types, when possible, of medical interventions.

There may come a time when our technological know-how will move beyond current genetic testing scenarios (both pre- and postnatal testing) and allow scientists to create children with enhanced traits, such as resistance to disease, increased strength, and enhanced memory. Furthermore, one day we may also be able to engineer purely aesthetic enhancements like hair color/texture and height. "Choosing Our Genes" confronts the visitor with these possibilities and asks the visitor to consider whether these types of changes are reasonable. Will they, for example, upstage the medical benefits of genomics? And will those who cannot afford enhancements be relegated to a genetic underclass? Similar ethical and social quandaries face scientists, policymakers, and the general public when considering a wide range of genomic-related technologies. In "Reshaping Our World," one of the final sections of the exhibit, the visitor can explore uses for several of these technologies, including genetically modified foods and cloning. Again, the ethical conundrums and real-life consequences of these genome-driven technologies are explored.

Another way to engage visitors is to gauge their opinions on genomic subjects. Polling stations were set up at several locations in the exhibit as a way to encourage participation and give visitors insight into their own genetic literacy. Sample questions included: "If records of genetic information were maintained by physicians, should employers be allowed to have access to those records?" and "If records of genetic information were maintained by physicians, should law enforcement agencies be allowed to have access to them?" The answers given by visitors are immediately compared to two larger databases: (1) answers of other visitors to "The Genomic Revolution" and (2) answers from a nationwide Harris poll commissioned by the museum. That poll was tremendously useful in establishing the exhibition team's understanding of the public's perceptions and misperceptions of the genome and also helped focus the exhibit on areas of public interest.

The contents of "The Genomic Revolution"—genetic enhancement technologies, gene therapy, genetically modified organisms, cloning, and the use of DNA in forensics and criminal justice—consistently conjure up visions of Aldous Huxley's *Brave New World*, originally published in 1932. Journalists and other popularizers of science may have overcited this prescient work to the point of cliché, but this does not detract from the lasting social importance of Huxley's vision. *Brave New World* was written and published during an important moment in the history of science. At the time the makers of biology's Modern Synthesis (the coupling of Charles Darwin's theory of evolution with mendelian genetics), including most prominently Huxley's brother Julian, ushered in a new era of biology. In this nascent science Huxley, himself a member of one of modern biology's "first families,"[8] saw potential for the creation of a new world order—a society not based on values of free will and democracy but a world in which men and women were to be "adapted and enslaved" to science, genetically engineered to carry out their station in life.[9] Today we seem much closer to the technological perversions of *Brave New World*. As Huxley himself said:

> *Brave New World* is a book about the future and, whatever its artistic or philosophical qualities, a book about the future can interest us only if its prophecies look as though they might conceivably come true.

Embryo selection, genetic engineering, and cloning—all contemporary technologies—both echo and heed *Brave New World*.

But it is not only Huxley's prediction of a dystopic future that interests us. It is also the role that science plays in the brave new world that worries us—that is, that science itself is a powerful narcotic. We like to think of this conception as "science as soma." In *Brave New World*, soma was the intoxicant and tranquilizer that the citizens of Utopia addictively consumed to blunt the pain of what Huxley called their "insane" lives. In our genomic era we worry that a popular understanding of genomics may become something like soma, dulling our collective craving for answers in increasingly complex times. In other words, genomics may capture our minds and numb our spirits, convincing people that their genes exercise final control over their individual and collective destinies. Despite the misconceptions, nothing could be farther from the truth. We must take care in making the genome public, so that the popular meaning of genomics is not reduced to something more powerful than the science it brings us. That is what the museum hopes will be the final achievement of "The Genomic Revolution."

Notes

1. Craig Venter is the president of Celera Genomics, a private biotechnology company. Francis Collins is director of the National Human Genome Research Institute at the National Institutes of Health. In separate efforts, both Venter and Collins completed first drafts of the human genome in June 2000. More recent studies by Stephens et al. using haplotype analysis have come to similar conclusions (J. Claiborne Stephens, Julie A. Schneider, Debra A. Tanguay et al., 2001, "Haplotype variation and linkage disequilibrium in 313 human genes," *Science* 293:489-493.)

2. Comments of C. Venter at the Gene Media Forum, July 20, 2000.

3. Daniel Kevles, 1985, *In the Name of Eugenics: Genetics and the Uses of Human Heredity,* Harvard University Press, Cambridge, MA.

4. It is an historical irony that the Second International Congress on Eugenics opened on September 22, 1921. Seventy-nine years later to the day another museum conference, "Sequencing the Human Genome: New Frontiers in Science and Technology," opened.

5. Charles B. Davenport, 1923, *Eugenics, Genetics and the Family: Volume 1, Scientific Papers of the Second International Congress of Eugenics,* Williams & Wilkins, Baltimore, MD.

6. Celeste Michelle Condit, 1999, T*he Meanings of the Gene: Public Debates About Human Heredity,* University of Wisconsin Press, Madison, WI.

7. Sharon Schmickle, *Star Tribune* (Minneapolis, MN), October 6, 1999, p. 9A.

8. Huxley's grandfather was Thomas Henry Huxley, one of the nineteenth century's most renowned zoologists and a close friend of Charles Darwin. Aldous's brother Julian, the architect of biology's modern synthesis, was an influential scientist like his famous grandfather. It was the potential of what could be that Huxley saw in the modern synthesis that permeates *Brave New World*.

9. "Foreword," 1989, *Brave New World,* reprint, Harper Perennial, New York, NY.

Part I

Genome Science and the New Frontier

Nicholas Wade

Introduction

 So powerful a body of knowledge is the human genome that it is surely likely to bring some problems along with its many benefits. There are three areas in which the consequences of genomic knowledge might be expected to give us pause. These relate to our view of human nature, to the genome as a means of human identity, and to the impending decision on whether or not to modify the genome.

The genome is the biological programming that defines the organism. We have two legs, arms, and eyes, and no horns, tail, or wings because that is what the human genome calls for. It is easy enough to accept that natural selection has defined in great detail the contours of our bodies. But has it done the same for our minds? To the extent that our minds too are important for our survival, we could expect them to have been just as strongly shaped by evolution and their operating rules written into our genes.

Few people like to think that their higher cognitive processes are under genetic control. But this is a point on which the genome may hold many surprises for us. If you look at primates' social behavior, at gorillas with their harems, chimpanzees with their multiple-male bands, humans with their close approximation to monogamy, it is hard not to suspect that the rules of our sociality, at least in broad outline, are written somewhere in the genome, just as they are in the genomes of other primates. The genes that govern our behavior are unlikely to determine every detail. We live in too complicated a society for preprogrammed behavior to be effective. It is more likely that the rules, such as they are, just set a general direction. It would be absurd to expect genes for speaking French or English or Japanese. But there surely are genes that lay the basis for grammar and genes that make it possible for children to acquire whatever language they hear spoken around them.

Another universal human behavior that takes different forms in different societies is religion. Is there a gene that causes a propensity to believe? If so, will we be happy when we find it? Biologist E. O. Wilson writes in his recent book, *Consilience*, that the human mind evolved to believe in gods; it did not evolve to believe in biology.[1]

We are shaped by a set of instructions that define our limits and maybe set our goals. There is no guarantee that everything we find in the genome will enhance our self-image. Human nature is a brew of strange elements, and we do not know yet what particular mix of murderousness and mercy has made us the sole survivors of the once-diverse hominid line.

The genome also bears strongly on human identity. It contains within it a whole series of different identity markers that track every individual's history from the ancestral human population to the present day. With one set of DNA markers—the microsatellites used by forensic laboratories—you can identify any individual in a population almost uniquely. With another set of markers, you can reach back a couple of centuries and estimate, say, that Thomas Jefferson was the probable father of the children born by his slave Sally Hemmings.

If you take another set of markers, you can reach back to the 10 sons of Adam and the 18 daughters of Eve, fanciful names for the major lineages that radiate from the ancestral human population. If you care to send a scraping of the cells from inside your cheek to a company called Oxford Ancestors and a check for $150, you can learn which of the 18 daughters of Eve your line belongs to.

All around the world, people are organized into social groups, centered around blood relationships, family or extended family, clan or tribe. These blood relationships are a surrogate for the information that is in the genome. People are intensely interested in their past and where they came from. When genome scans become cheap and routine, this genealogical information will be available in more copious form than ever before. It will place everyone who cares to know on a great family tree, with a single trunk and branches corresponding to the world's major ethnic groups. Will that genetic tree prove to be healing or divisive? If you look at the trunk, we are indeed one family of very recent origin, perhaps as little as 50,000 years. But if you look at the twigs, we are many separate clans and cultures.

The third aspect that may give us problems in the future is germ-line engineering. We strive hard to build a just society, but we ignore a glaring source of inequality. The fact is that in the lottery of conception some of us are dealt good genes and some bad. Until now, there has been nothing we could do to help parents improve their children's genetic endowment. With the sequencing of the genome, we are being propelled ever faster to a decision that was perhaps made inevitable by the beginning of modern genetics, that is—will we change the human genome for the better?

Germ-line engineering is not a subject scientists generally enjoy discussing in public. It is premature; everyone agrees that we do not possess the technology or the wisdom to do it yet, and it upsets people for a variety of reasons, some of them well founded. Still, the genome is going to thrust this debate on us sooner or later and the better the public understands the issues, the better its decision is likely to be.

There is no unanimity within the scientific community. The biologist E. O. Wilson, for example, is against it. He calls the idea of genome engineering "the most profound intellectual and ethical choice humanity has ever faced. Our childhood having ended, we will hear the true voice of Mephistopheles," he writes in *Consilience*. He believes that to rid the genome of its apparent imperfections in favor of pure rationality will be to "create badly-constructed, protein-based computers. It would lead to the domestication of the human species. We would turn ourselves into lap dogs."[2]

Wilson thinks those imperfections of character are essential to our nature. But James Watson, codiscoverer of DNA, talking principally of imperfections in health, is eager to harness our genetic knowledge for human benefit. At a recent conference, he said

> the biggest ethical problem we have is not using our knowledge, people not having the guts to go ahead and try and help someone. We're always going to have to make changes. Societies thrive when they are optimistic, not pessimistic. And another thing, because no one has the guts to say it, if we could make better human beings by knowing how to add genes, why shouldn't we do it? What's wrong with it? Who's telling us not to do it? We should be honest and say that we shouldn't just accept things that are incurable. I think what would make someone's life better, and if we can help without too much risk, we've got to go ahead and not worry whether we're going to offend some fundamentalist from Tulsa, Oklahoma.

So there you have it: two quite different views from two very thoughtful people.

Whether or not we take Wilson's or Watson's path will be a hard choice for society to make. It will learn, maybe, that genomic technology is neither good or bad in itself but can have good or disastrous consequences depending on the wisdom with which society shapes it. In the following section four distinguished scientists explain how these genomic techniques have developed and where they are likely to lead. Harold Varmus, director of the Memorial Sloan Kettering Cancer Research Institute, introduces the reader to the science behind the genome and discusses the effects of genomics on science and on health care. Eric Green, director of the NIH Intramural Sequencing Center, explains, step by step, how the human genome is actually sequenced. J. Craig Venter, former President of Celera Genomics, describes the whole genome shotgun sequencing approach that he utilized to complete a rough draft of the human genetic code. Finally, Leroy Hood, whose advances in sequencing technologies helped to bring about the genomic age, writes about the future of gene sequencing technologies.

Notes

1. Edward O. Wilson, 1999, *Consilience: The Unity of Knowledge*, Random House, New York, NY.
2. Ibid., p. 157.

Harold Varmus

What Does Knowing About Genomes Mean for Science and Society?

Genomics is an accelerating and complex step in the longer history of molecular biology and genetics. It has become an integral and essential element of biotechnology and molecular biology, transforming the very way in which modern biology is conducted. On a broader scale, genomics is forging new perceptions of how life works and changing our concept of our world and our origins. On a more practical and immediate level, the field of genomics is already affecting our lives—for example, in the food we eat, the law we practice, and the medical care we receive. The changes we are now witnessing in our daily lives grew out of the recent history of cloning genes, extracting their secrets, and using their products through biotechnology.

We have just lived through what many of us think of as the

20

Century of the Gene, which importantly is also characterized by discoveries in physics and computer sciences. The Century of the Gene
is conveniently demarcated by the rediscovery of the principles of
Mendel in 1900, by the description of DNA as a double helix at
midcentury (1953 to be exact), and in 2000 by the announcement of
the rough draft sequence of the full human genome. There is much
excitement about the genomic revolution. Thus, it is essential that
the public understands its meaning and implications because we, as
a society, will face numerous choices about its applications in the
coming decades.

Biology 101

DNA is basically a chemical—a long chain of units called nucleotides,
with four types distinguished by components called bases, specifically adenine, thymine, cytosine, and guanine (A, T, C, and G). Pairs
of these nucleotides (A always with T, and C always with G) make
the twisted chain known as the "double helix," or DNA. Simply put,
DNA is a very long chain made up of these four basic chemical
substances.

In the simplest terms, DNA makes up genes, which make up
chromosomes, which together comprise the genome. Thus, a *gene* is
a functional unit embedded in the long chain of DNA (see Figure 1).
It is a functional component comprised of hundreds or thousands of
base pairs of A, T, C, and G. A *chromosome* is a long piece of DNA that
might contain one or thousands of genes. A *genome* is an entire collection of DNA—that is, all the chromosomes of a single organism
that comprise its complete repertoire of genes. Humans have 23 pairs
of chromosomes; many simple organisms, like bacteria, have just
one pair.

In this century, remarkable progress was made when scientists
moved away from the classical concept of a gene as an instrument
that a cell inherits, accounting for variations in the behavior and
appearance of individual organisms, toward a modern molecular
concept in which a gene is a physical thing. This concept—that a

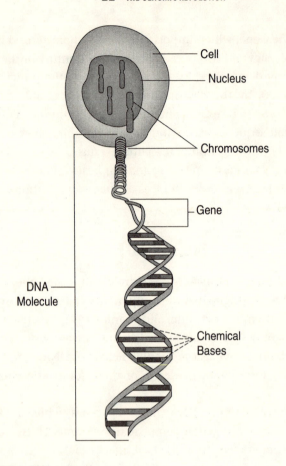

FIGURE 1 DNA molecule.

gene is something embodied in a sequence of DNA—has produced the greatest reverberations for the coming century of biology.

One of the most important contributions made by those who studied organisms in the twentieth century was the development of the central dogma of biology, which can be compared to basic computer technology (see Figure 2). Biological information, which is embedded in a DNA sequence, can be compared to the hard drive of a computer. This information is then read out in the cell into a much more labile form, called RNA or a messenger, much like a floppy

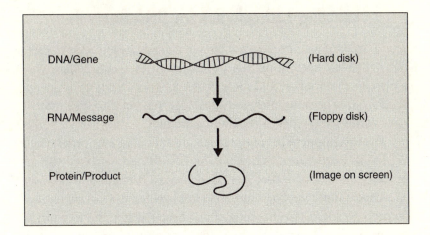

FIGURE 2 Hierarchy of biological information.

disk. From that form the cell is able to extract the information encoded in the nucleotide sequence to make what is in most cases the most essential, most important product of a gene—namely a protein. And that could be analogized to the image on a computer screen.

In thinking about the potential similarities and differences of different organisms, the size of the genome is impressive. Science has known for at least 50 years roughly how big different genomes are. As one might expect, organisms that may seem relatively small, like viruses and bacteria, tend to have relatively small genomes— perhaps 3 million base pairs or, for most viruses, much less. As genomes get larger, the organisms seem more complicated and even more interesting. For example, yeast has about 12 million base pairs, while humans and other mammals have roughly 3 billion base pairs. The size of the whole genome, however, is not necessarily directly reflective of the number of genes. Indeed, as the number of nucleotides goes up, the number of genes seems to go up at a much reduced pace.

Enabling Technologies for DNA Analysis

The analysis of genomes to find out what genes are encoded in them has depended heavily on the development of technologies. New tools can manipulate and clone DNA and subject it to sequence analysis. It is because of these technological breakthroughs that we have witnessed the development of genomics.

The first moment of culmination came about in 1995 when Craig Venter and colleagues presented the first full sequence of the genome of an organism that was free-living—a bacterium as opposed to a virus. (Many viral genomes had been sequenced over the previous two decades.) The mapping of this bacterial genome was critical because suddenly we could see in one large multicolor map a picture of the full repertoire of a genome, highlighting genes from different functional classes in different colors. By viewing the map it is possible to, for instance, see how genes for different functions are arranged throughout the chromosomes.

A yet more imposing accomplishment occurred in 1996 when an international consortium of laboratories determined the first complete sequence of a free-living, eukaryotic genome. This organism, a yeast, contains a nucleus with 16 chromosomes, representing over 12 million base pairs and roughly 6,100 genes.

Over the past couple of years, genome maps of two important experimental multicellular organisms were unveiled with great excitement: the roundworm *Caenorhabditis elegans* with six chromosomes and 19,000 genes, and *Drosophila melanogaster*, or fruit fly, with four chromosomes and 13,000 genes. And, of course, we now have a rough draft of the human genome, the results of a mainly federally funded public effort and one in the private sector, from the company Celera Genomics.

The Metaphors Game

As genomics has entered the public consciousness, scientists and journalists have engaged in an entertaining activity I call the metaphors game, which has provided some important first efforts to explain what it means to have analyzed a genome. One of the most common metaphors is the "Book of Life," which conveys the notion that genetic information is written down in our genomes. The task of the cell is to read that information, putting it into the slightly different language of RNA. Then the information embedded in the RNA is read out by protein synthetic machinery called ribosomes.

The metaphor is also useful with respect to size. We all know what books feel like and how much time it takes to read them. If we assume there are 3,000 letters on a single page of a book, it would take 1,000 volumes each of 1,000 pages to represent all the information present in an individual's genome.

Another useful metaphor is the blueprint because it indicates the notion of instructions for building or making something. One of the things that is particularly fascinating about genomes, especially genomes of human beings, is that they contain in one single cell all of the information needed to make a human being. So in a sense the blueprint of life is contained in that first single cell, the fertilized egg. We end up being compilations of 10^{13} cells, many of which have very different functions and all of which were made in response to a single set of instructions.

The metaphor of a map also serves us well. Indeed, we began to do genomics by making maps not of entire nucleotide sequences but of positions of genes and physical entities, such as sites at which certain enzymes recognize and cut DNA. The map also implies discovery. Early discoverers do the work of laying out the physical domains of the genome, while others slowly and methodically fill in the details.

The periodic table, well known to chemists, provides another metaphor for the genome. Eric Lander, of the Massachusetts Institute of Technology, may have been the first to draw this comparison. It is useful not so much because it exactly corresponds to how we use genetic information—the analogy with chemicals is not perhaps precise—but because elucidation of the periodic table is ongoing. Although it was first described over 100 years ago, new elements are still being added and new ways of using the elements are being devised. The first rough draft of the periodic elements signaled not the end but the beginning of chemistry. Similarly, the first draft of the human genome tells us that we have simply reached the beginning of a new understanding of living organisms.

A metaphor I tend to favor is that of a machine. One of the things that is clear from the results of the genome project is that we now have at least the instructions for making all the parts of a cell. The challenge will be taking that cell's machinery apart and learning to put it back together.

Tools for Understanding the Genome

What are the technologies that will allow us to gain an understanding that goes beyond just the raw sequence of the genome to a fuller comprehension of biological systems? One of the challenges of genomics is to move from detailed sets of data that illustrate the positions of nucleotides in some random piece of DNA sequence to a form that can be read more easily. To interpret these long strings of data, we must write the data down in digital script on a hard disk. Instrumentation, robotics, 96-lane sequencing devices, and computer hardware and software for storage and interpretation retrieval are essential to this interpretation.

Genomics is not just a biological science; it is a science based on engineering and computer science. There would be no genomics without the ability to store, compare, analyze, search, and annotate all of the sequences generated in the genomic age. The first goal of genomics is to try to understand the raw sequence, which is as meaningless to me as it is to you without some kind of interpretive component. Fortunately, our current challenge is markedly aided by the existence of what we call the tools of bioinformatics. These tools are not just simple personal computers but very ingenious kinds of software that allow analysis of a sequence, followed by efforts to elicit from that analysis missing data—that is, information not available from experimental data.

In my first representation of a gene, I gave you a very simplistic notion of just DNA giving rise to proteins, but of course life is more complicated. As we have learned more over the past 30 or 40 years about the physical nature of a gene, we have come to understand that a gene is actually a fairly complex element. First, there are the sequences that directly encode proteins: each set of three nucleotides makes a word that spells out for the ribosome an amino acid to be inserted in the protein chain. But in addition to the information that encodes proteins, which are embedded in elements of genes called exons, there are sequences in between, called introns, which are not translated into proteins. Moreover, there are sequences in front of the gene and in the middle of the gene that determine when and where that gene will actually be read to make a protein product; this is called regulatory information. In addition, there are sequences farther downstream that may influence where the reading process is terminated and sequences in between that govern the processing of the RNA as it goes through its machinations that lead to the production of the proteins (see Figure 3).

The difficulty in trying to recognize the exon sequences that encode proteins has given rise to a problem in analysis—namely, that it is not yet clear how many genes there are in the human genome. By extrapolation one might have guessed a number in the range of 60,000 to 80,000. But as of late 2000 estimates of 30,000 seemed more likely. What this reflects is considerable uncertainty about how to interpret the sequences and the need for a prolonged annotation process following the production of the raw sequence.

Fortunately, we have a method to assist us in trying to recognize genes beyond some of the simple signals that denote whether or not a DNA region is likely to encode protein. This method is made possible by the fact that we now have the sequences of many genomes and can compare sequences that reappear. One of the great messages of genomics is that we—plants, insects, bacteria, and all animals— are all related. As we compare genes we find a surprising number in different organisms that have homologues in other species. In the analysis of the yeast, fly, and worm genomes, it became increasingly apparent that, as we looked at sets of sequences, the number that are unrecognizable diminished quite rapidly (see Figure 4). These comparisons provide a clearer notion of how genes evolved and facilitate recognition of gene functions.

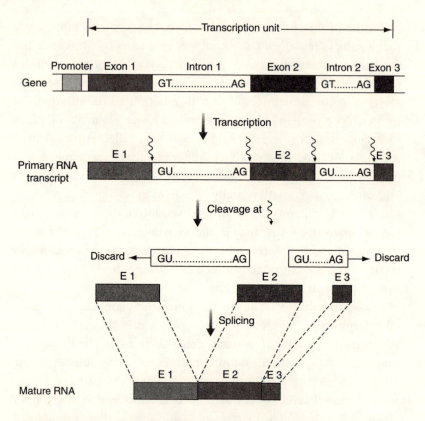

FIGURE 3 RNA splicing involves endonucleolytic cleavage and removal of intronic RNA segments and splicing of exonic RNA segments.

There is no simple linear correspondence between the size of a genome and the number of genes. This might mean that indeed some of the sequences that are in between genes or that interrupt the coding sequences of genes are going to be of extreme importance in understanding how human beings came to be a lot more complex than roundworms. In addition, it is important to emphasize that a single gene may add a lot more diversity to a cell's function than one might expect. This is so because each gene can be read into multiple RNAs by starting and stopping at different points and gluing the pieces of RNA together in different combinations, through a phenomenon called splicing. Each series of events can generate a unique

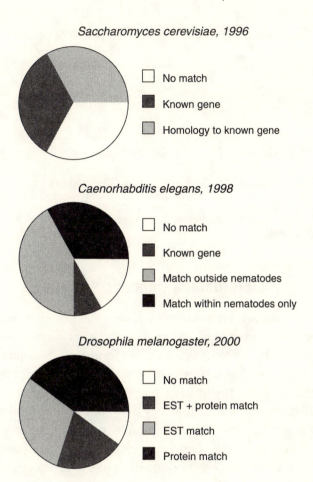

FIGURE 4 Comparison of worm, fly, and yeast genomes.

coding sequence, meaning that each gene can encode multiple related proteins. In addition, each RNA can encode different proteins by starting and stopping in different places and even changing the frame in which the reading is done. Each protein can also be modified by a number of processes—for example, by cleavage into multiple proteins or by the addition of other chemical groups, like sugars and phosphates that change the function of the protein. This adds a very deep level of complexity to the situation (see Figure 5).

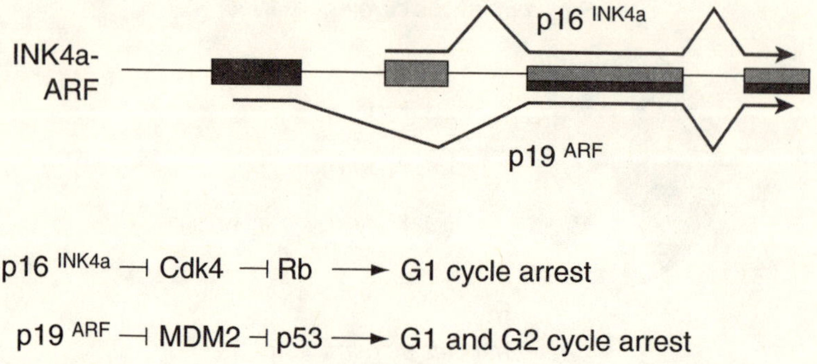

p16 ^{INK4a} ⊣ Cdk4 ⊣ Rb ⟶ G1 cycle arrest

p19 ^{ARF} ⊣ MDM2 ⊣ p53 ⟶ G1 and G2 cycle arrest

FIGURE 5 Overlapping protein-coding sequences (genes) in a region of mammalian DNA.

Understanding Genomic Variation

Despite the remarkable similarities among the genomes of various organisms, genomics also informs our notion of genetic variation. Genomes can be different by variations in nucleotide sequence but also through duplications or deletions of DNA, through combinations that rearrange the order of genes, or by insertions of so-called moveable elements of DNA that may have come from viruses or from DNA that has its own "engines" to drive its movement around the genome. And, of course, there is the process of sexual reproduction that generates new combinations of genes. The effects of these changes may be the differences we see among species—namely evolution. These changes account for differences among individuals and may also account for changes that occur during the lifetime of an individual. For cancer biologists like myself, this last possibility is crucial because cancer is a disease of genetic change, much of which occurs in somatic cells during the lifetime of an individual.

What we are now equipped to do through genomics is to look for variations by starting with the genome itself, ignoring for the moment the variations in appearance and behavior that result. By analyzing genomes to look for simple genetic variations, or single nucleotide polymorphisms (SNPs), we then face the task of identifying which of those variations are important for interpreting the differences in behavior, in the appearance of disease, in tolerance to drugs, or many other traits. Although we all share 99.9 percent of our genome with the rest of the human race, each of us differs at roughly 1 in every 1,000 nucleotides from each other. While these are small differences, the genome is big; thus, there are about 3 million differences between my genome and the genome of my neighbor. We would like to know what the significance of those differences might be. Thus, there are many kinds of clues to gene functions provided by inspecting gene sequences, by looking for gene families and homologues in other species, and by studying genetic variation.

Effects of Genomics on Basic Science

One way to perceive how a new advance is affecting the conduct of science is to look at advertisements in scientific journals. A recent advertisement in the magazine *Science* summed up the complexity of genomics with this slogan: "So many genes, so little time."

Most of us who work in experimental biology are used to working with a small number of genes, possibly 1 to 10. Suddenly we are confronted with the possibility of working with 30,000 genes at a time. This creates a whole new set of difficulties that require technological advances and new ways to try to interpret biological information.

Nanotechnology has provided us with a wonderful invention called the gene chip. Basically this is small-instrument technology to generate devices that allow more numerous and more rapid analyses—for example, DNA sequencing without having to use a conventional sequencing apparatus to seek out mutations. Similar kinds of chips are available for looking at gene expression. A robotic instrument can apply to a single glass microscope slide a representative piece of every one of the 6,100 genes present in the yeast genome. This allows analysis of each of those genes—for example, as the organism goes through the cell growth cycle under different metabolic conditions to try to understand how genes are regulated and what different genes might do.

These technologies provide us with an incredibly complex array of information, so fraught with ups and downs and interactions that most of us simply do not have the mental capacity to try to make sense of these large datasets. Efforts to try to search out the influence of any single protein in this complex array is akin to trying to figure out how different transistors might be operating in complex electrical circuitry. So we are challenged in a way that requires biologists to seek the help of mathematicians, engineers, and other folks who have rather different approaches to science to try to make sense of information that has been gathered in what are basically traditional experiments in cell and molecular biology. These technologies allow us to address the behaviors of tens of thousands of genes at once.

Applications of Genomics

Increasingly, the criminal justice system is permitting the use of DNA analysis to determine the guilt or innocence of crime suspects. These analyses, using specific genetic markers that indicate variations in human sequence, are also being applied in history, anthropology, linguistics, and many other fields in efforts to understand human origins and identify different groups based on genetic information.

Two additional fields in which genomics has had pronounced effects in recent times are the food industry and medicine. The advent of genomics is going to markedly increase the repertoire of possible genes to transfer from one organism to another—for example, to change the nature of agricultural products, both plant and animal in origin.

Many applications in medicine are already in practice. Techniques for assessing genetic risk of disease, now available for some diseases that occur largely as the result of a single mutation in a single gene, are being expanded to efforts to understand diseases that result from a combination of mutations. Analysis of SNPs and other variations among human individuals will become important to improving assessments of genetic risk. One important aspect of this, particularly in the context of cancer and infectious diseases, is to understand disease processes by having all the genes at our fingertips and being able to look at gene behavior during the development of disease.

Genomics is promoting a tremendous interest in novel therapies of which gene therapies will be only a minority. Therapies based on disease mechanisms will be better understood using these new tools of molecular biology. Chip technology is already changing the way we think about the diagnosis of cancer. For example, examination of roughly 19,000 genes that are turned off and on in samples from lymphoma patients allows us to separate patients (who might otherwise seem indistinguishable by conventional criteria) into groups that may respond differently to therapy and have very different outcomes. This refinement of diagnosis is something that depends heavily on the techniques arising as a consequence of our understanding of genomes.

But things do not always occur that quickly. To illustrate, in the year 2000 we celebrated a new drug called STI-571, or Gleevec, that appears to be effective in the treatment of chronic myeloid leukemia. The development of that drug is based on a genetic understanding of the disease that can be traced back to the discovery of an abnormal chromosome 40 years ago. It is worth remembering that, while genomics has tremendous power to allow us to look at all those genes that could affect the development of cancer and other disorders, we have to understand through conventional and laborious biological technologies the way in which those genes operate, their protein products, and then go through the process of identifying drugs that interfere with that process. The drugs then must be tested in the clinic.

Effects of Genomics on Individuals and Society

We all receive a set of chromosomes—23 from our father and 23 from our mother—that constitute our genome. Although we are a reflection of our genetic inheritance, we are more than our genes. I do not know any biologists who believe in genetic determinism—

the idea that what we are is only represented in our genes. Genes are only one component of who we are. But clearly the sense that genes are an important component of who we are is both edifying and alarming.

There has been tremendous concern about consumer interests and how they are affected by the genetics revolution. Many of us are concerned about what could be done with genetic information or about how it might be interpreted. It is disturbing that this information might not be kept in a private fashion or might be misused to discriminate by employers, insurers, or others.

We also have to remember that the risks inherent to genomics are just one side of the coin. Genomics includes technical innovations that could generate food for Africa, produce medicine for individuals who are ill, and provide a deeper understanding of science in general, including our origins and the nature of our variations. These benefits are a counterweight to the concerns voiced about the misuse of genetic information. Indeed, one of the purposes of this book is to reflect on the fact that science is a human instrument, one that can be used wisely or foolishly. It is only an informed public that will decide what will happen.

Eric Green

Sequencing the Human Genome

Elucidating Our Genetic Blueprint

A central rationale for the Human Genome Project is that a complete working knowledge of our genome will provide critical information and tools for advancing our understanding about the genetic basis of human health and disease.

In the cell's nucleus is our genetic blueprint, housed in structures known as chromosomes. Chromosomes contain our DNA, which consists of a four-letter alphabet of chemicals, specifically A (adenine), T (thymine), C (cytosine), and G (guanine). It is the sequence of these chemicals that encodes the information contained in our genetic blueprint. In 1953, James Watson and Francis Crick published a now-famous paper in *Nature* reporting that DNA has a double-helical structure and is the molecule of heredity in all free-living organisms, a research accomplishment that earned them a Nobel prize.

Since 1953 there has been a virtual explosion in the amount of information about the structure and function of DNA. This has led to spectacular advances in genetics, fostering the molecular biology revolution of the late 1970s and early 1980s and ushering in the era of DNA cloning and powerful new tools to study biology. However, this just set the stage for the genomic revolution of the 1990s. At the centerpiece of this revolution is the Human Genome Project, which celebrated its decade mark in the year 2000.

The Role of Model Organisms

An initial point worth stressing about the Human Genome Project is that its name is somewhat of a misnomer since it is not limited to the study of human DNA. Rather, under priority study are organisms such as a yeast, a fly, a worm, the mouse, and the human. There are many reasons for inclusion of these so-called model organisms in the Human Genome Project. First, they all share the same fundamental DNA structure and many of the same genes as the human genome. Second, they have been studied and genetically manipulated in laboratories for decades. And third, they are at distinct evolutionary points, having separated from humans 80 million to 1,000 million years ago (see Figure 1). A key goal of the Human Genome Project has been to characterize the genomes of these organisms, thereby providing information about the genetic blueprints responsible for unicellular, multicellular, mammalian, and human biology.

Reading the Book of Life

A genome can be thought of as an encyclopedia set of books, each containing the precise order of our Gs, As, Ts, and Cs, as if typed out on pages of those books. The sizes of the genomes of various organisms vary. The human genome contains about 3 billion bases, or 3,000 megabases. The genomes of the fruit fly, worm, yeast, and bacteria are all substantially smaller than the human and mouse genomes (see Figure 2).

The 24 human chromosomes (1 through 22, X, and Y) can be

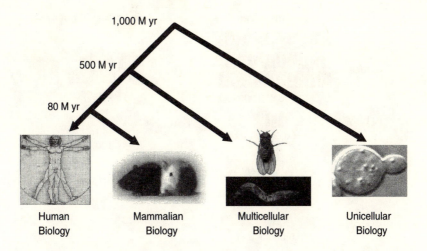

1,000 M yr

500 M yr

80 M yr

Human
Biology

Mammalian
Biology

Multicellular
Biology

Unicellular
Biology

FIGURE 1 Many model organisms are also being studied as part of the Human Genome Project, including a yeast, a fly, a worm, and the mouse. Representing various stages of organismal development throughout history, the DNA of these unicellular, multicellular, and mammalian organisms provides clues to the significance and function of human DNA.

visualized microscopically, revealing the cytogenetic map of the human genome—something established long before the Human Genome Project began (see Figure 3). This map provides a useful framework for analyzing the DNA in each chromosome. In fact, one can think of the human genome as a 24-volume encyclopedia set.

The Human Genome Project sought to elucidate the human genetic blueprint in two stages. In the first stage the DNA from each chromosome has been studied and organized by a process called mapping. In the second stage the sequence of the organized DNA has been determined, or read. In the first physical mapping stage, each starting chromosome is broken up into pieces and the pieces are recovered as DNA clones. These clones are then characterized to determine which ones have DNA in common. Based on this information, the clones can be overlapped or arranged relative to one another and organized into what are known as "contigs," which are collections of clones that together contain a contiguous segment of

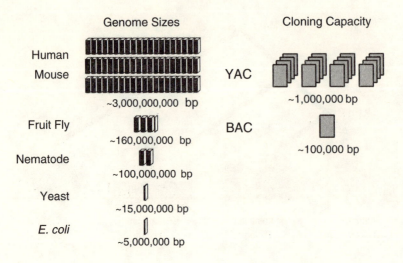

FIGURE 2 The relative sizes of genomes of various organisms. Each book contains 50 million bases: the human genome contains about three billion bases, or 3,000 megabases. The genomes of the fruit fly, worm, yeast, and bacteria are all substantially smaller than the human and mouse genomes, which, like all mammals, are roughly the same size.

the starting DNA. This activity is highly analogous to putting together a jigsaw puzzle.

Two cloning systems became instrumental to the Human Genome Project's effort to map the human genome. In one case, larger pieces of the genome puzzle were cloned as artificial chromosomes in yeast, known as *YACs*. Later, another cloning system became available in which smaller pieces of DNA could be cloned as artificial chromosomes in bacteria, or *BACs*. In essence, each YAC contains many pages or roughly chapter-sized pieces of cloned DNA, while BACs provide individual pieces of the puzzle, roughly page-sized pieces of the human genetic blueprint. Because of their larger size, YACs played a dominant role in constructing the first-generation physical maps of human chromosomes.

From about 1990 to 1997, a central activity of the Human Genome Project involved constructing physical maps of the 24-volume human encyclopedia set. This was done chromosome by chromosome using YACs, thereby providing chapter-by-chapter ordering

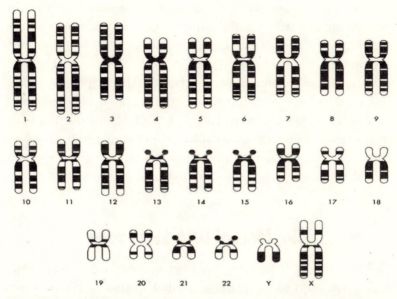

FIGURE 3 Cytogenetic map of the human genome. The human genome consists of 23 pairs of chromosomes: 22 paired autosomes and one pair of sex chromosomes—two X chromosomes in females and one X and one Y chromosome in males.

across every volume of each human chromosome. For various reasons it turned out that YACs are not well suited for sequencing the human genome. In contrast, the smaller BACs are more appropriate. And so a major emphasis from 1998 to 1999 was construction of page-by-page maps of each human chromosome using BACs. These were the second-generation physical maps of the human genome. With maps like this becoming available across the human genome, it was then time to start reading the pages—that is, taking each BAC-sized piece of DNA and determining the precise order of its roughly 100,000 bases.

Whose Genome Are We Sequencing Anyway?

Fundamentally, the DNA of any two humans is approximately 99.9 percent identical. For the first pass of establishing the human genome sequence, it really does not matter whose DNA was selected because we are just determining a reference sequence that will provide an infrastructure for future studies. In addition, any one individual is actually a mixture, or mosaic, of DNA, half from each parent.

The human genome sequence established by the publicly funded Human Genome Project was generated in a reasonably careful way, from appropriate materials and libraries. The BAC libraries were made from a series of individuals, completely anonymous. What has been sequenced and what exists on the Internet as the human sequence should be viewed as a fictitious individual—a hypothetical mosaic sequence.

The Tools of DNA Sequencing

The field of DNA sequencing has a rich and exciting history, with a major crescendo taking place in the last quarter of the twentieth century. A key contribution was made in 1977, when Fred Sanger described a new method for DNA sequencing, the fundamental basis of which is still being used at the present time and primarily has been responsible for sequencing the human genome. Since that time there have been a number of evolutionary improvements in DNA sequencing, including some key advances made in sequence automation by Leroy Hood's group in the 1980s.

Numerous incremental advances in DNA sequencing took place in the 1990s as well. The net effect has been to substantially increase the efficiency of DNA sequencing, such that a single person working in a sequencing laboratory can now generate upward of a million bases of sequence in a given year, up almost three orders of magnitude from where it was just a quarter century ago (see Figure 4).

The Sanger method for DNA sequencing involves the termination of newly synthesized DNA molecules with a specially modified base (G, A, T, or C), which essentially marks the position of that base in the starting template. For many years the standard approach involved tagging the DNA molecules with radioactivity, separating the DNA by gel electrophoresis, and then detecting the radioactive DNA by exposure to X-ray film.

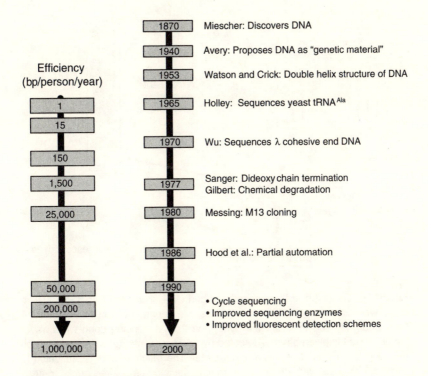

**Efficiency
(bp/person/year)**

1	
15	
150	
1,500	
25,000	
50,000	
200,000	
1,000,000	

1870	Miescher: Discovers DNA
1940	Avery: Proposes DNA as "genetic material"
1953	Watson and Crick: Double helix structure of DNA
1965	Holley: Sequences yeast tRNA[Ala]
1970	Wu: Sequences λ cohesive end DNA
1977	Sanger: Dideoxy chain termination Gilbert: Chemical degradation
1980	Messing: M13 cloning
1986	Hood et al.: Partial automation
1990	
	• Cycle sequencing • Improved sequencing enzymes • Improved fluorescent detection schemes
2000	

FIGURE 4 History of DNA sequencing.

For many years the only way to generate large amounts of DNA sequence was to hire a lot of people to perform radioactive sequencing. Needless to say, this was very labor intensive, costly, and inefficient. This all changed with the development of automated fluorescent-based DNA sequencing methods. Specifically, these involve the use of fluorescent dyes that are tagged to the DNA molecule, with each color reflecting a different base. All reaction products can be loaded together and then separated electrophoretically (see Figure 5).

An exciting development in DNA sequencing in the past few years has been the introduction of new-generation sequencing instruments, which have some significant technical advantages compared to their predecessors. The impressive thing about this latest set

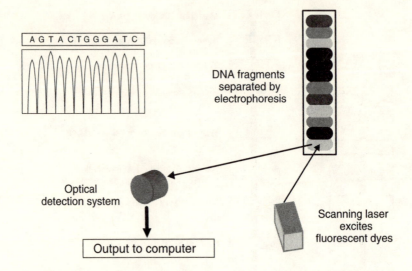

FIGURE 5 Detection of fluorescently tagged DNA. The DNA fragments are labeled with different color fluorescent dyes and separated from one another. A laser beam is used to excite the dyes, and then depending upon what wavelength of light is given off, it is detected by an appropriate optical system, which then communicates all of this to a computer. The resulting data is subjected to computer analysis, with the final information depicted as colored peaks, reflecting the base at that position in the starting DNA.

of machines is that they are extremely automated and therefore capable of high-throughput data production. For example, one machine can run unattended for at least 24 hours during which time it can analyze more than 1,000 samples, each providing about 500 to 700 bases of new DNA sequence.

Establishing the Human Genome Sequence

In short, the methods and instruments available for DNA sequencing have been remarkably refined over the past five years as part of the Human Genome Project. They now provide the ability to obtain large amounts of DNA sequence in sentence-sized segments in a very efficient fashion.

So how does one go from a page of one of those books, that is, a BAC clone, to actually determining the sequence of the 100,000 bases on that page? The most common way this is done is by a strategy known as "shotgun sequencing." In shotgun sequencing each individual BAC is taken, and large amounts of DNA are prepared from that BAC, as if one were simply photocopying that page of the book. The DNA is then randomly fragmented, as if the copied pages were put in a paper shredder (see Figure 6). The resulting fragments are then subcloned into a suitable vector. Thousands and thousands of resulting subclones are picked at random from that page, and sequence-sized reads are obtained from each. Large numbers of such sequence reads are generated from many places across the starting BAC. This provides highly redundant sequence information, which is then analyzed by a specialized computer program that assembles the sequence into sequence "contigs." At an early stage, prior to perfecting the generated sequence, the product is referred to as a

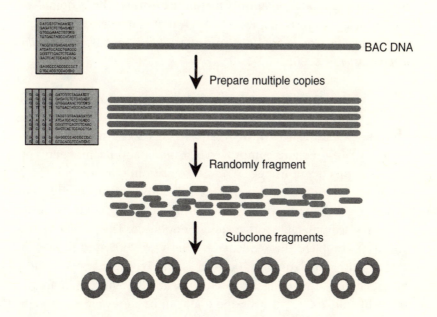

FIGURE 6 Subclone construction.

"working draft sequence." The process of polishing a working draft sequence to produce a final product is called "sequence finishing." This involves getting additional sequence reads to improve the accuracy of weak areas and filling in the missing sequences. The final product is a high-accuracy finished sequence.

A key feature of the publicly funded Human Genome Project is that all sequence data are made available every night, freely and openly, on the World Wide Web—including working draft and finished sequence. Thus, all new sequence data are immediately available to anybody around the world with an Internet connection. Even the groups that are generating the sequence do not have any prior or private access to their own data. This has not been the case for private efforts to sequence the human genome.

Shotgun sequencing using automated instruments has played a central role in the Human Genome Project. Indeed, the fundamental approaches used for sequencing the human genome were actually developed and carefully refined by first sequencing the smaller genomes of the model organisms mentioned earlier: the genome sequence of a yeast was completed in 1997, that of the nematode worm *Caenorhabditis elegans* in 1998, and that of the fruit fly *Drosophila melanogaster* in 2000.

Acceleration and Automation

The numerous advances in DNA sequencing that occurred throughout the 1990s—including improved instrumentation, refinements of shotgun sequencing strategies, and insight from sequencing model organisms' genomes—greatly heightened motivation and excitement to accelerate the pace for sequencing the human genome. As a result, a revised timetable was created for doing so. The previous plan aimed to complete the sequence by the year 2005; the new plan called for its completion by 2003 (see Figure 7). The new plan called for generating a working draft sequence for the entire human genome by mid-2000. This working draft can be compared to a rough draft of a manuscript, essentially containing all of the information needed but still requiring some hard polishing before completion.

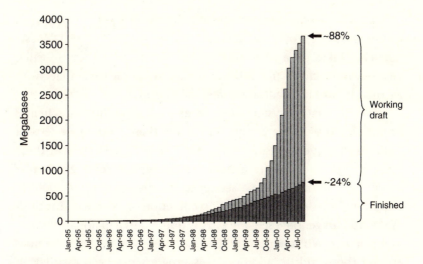

FIGURE 7 Timetable of human genome sequencing. The status of human genome sequencing, showing the amount of human sequence that has been generated in recent years. Particularly note the large amount of sequence that was generated in 1999 and 2000. Almost one-quarter of the human genome sequence is finished. The amount of working draft sequence, combined with the finished sequence, now totals roughly 90 percent.

Once a working draft sequence is generated, the subsequent years can be spent completing the project—that is, refining the sequence and advancing it from a working draft stage to a complete, highly accurate product. To accomplish this ambitious endeavor in such a short timetable, most of the effort was consolidated to a small number of sequencing centers around the world. Three of the five are supported by the National Institutes of Health (NIH), one by the U.S. Department of Energy, and one by the Wellcome Trust in England. These five centers, affectionately referred to as the "G-5," are responsible for generating about 85 percent of the human genome sequence. An additional 12 smaller groups from a number of different countries are responsible for generating the remaining human sequence.

The production demands at these five centers have been high, in some cases requiring the centers to accomplish more than a 20-fold increase in the scale of their operations. Automation has become a key characteristic of these centers. The entire process of DNA sequencing takes on a very industrial flavor. In fact, these research facilities more closely resemble factories producing cars or electronics than typical biomedical research laboratories.

The automation systems and the associated production schemes implemented by these groups now allow them to produce tens of millions of sequence reads every year. The dedicated work of these sequencing centers has resulted in the generation of a spectacular amount of human genome sequence. In the summer of 2000 an important milestone was reached: completion of a first working draft of the human genome sequence. This resulted in a major announcement at the White House and simultaneously at several venues around the world. In addition, this progress has been charted on various websites. For example, there is an NIH website where one can follow each chromosome as it is sequenced, which vividly illustrates how rapidly most of the human genome sequence has been generated. Two chromosomes, 21 and 22, are both finished, providing us the first glimpses into the complete genetic landscape of a human chromosome.

The Future

With such rapid generation of the human genome sequence, the challenge becomes learning how to assimilate all of these new data. The next key phase of the Human Genome Project is going to be the interpretation phase, analyzing all of the new sequence data and trying to figure out what the data mean. One can think of this as a rather challenging puzzle of 3 billion newly discovered pieces. Remarkably, the entire sequence of the 3 billion bases will fit on one CD-ROM, which underestimates its complexity.

The series of letters that make up our sequence—As, Cs, Gs, and Ts—are not in upper or lower case, there are no spaces between them, and there are no punctuation marks. Thus, in parallel with all of these efforts to map and sequence the human genome is the emergence of programs for analyzing this massive amount of sequence data. For the next several decades, a major priority will be to study these precise strings of letters and develop increasingly powerful computational and experimental methods for interpreting the sequence and identifying the relevant features. It took an ambitious endeavor to elucidate the 3 billion bases of the human sequence, and it will take a similar effort to eventually interpret it.

In summary, the scientific tools of discovery are being applied to elucidate the human genetic blueprint and to provide a powerful new infrastructure for all of biological research, including the study of human biology in health and disease. Many of us who have been immersed in the Human Genome Project believe that constructing maps and generating the sequence of the human genome and other genomes will bring about a revolution in the biomedical sciences. We will for certain look back at the year 2000 as a key turning point, and we will think about how we did research before we had the sequence in hand and how it all radically changed after the sequence was established. The most exciting developments of the genomic revolution are likely still to come, so stayed tuned.

J. Craig Venter

Whole-Genome Shotgun Sequencing

Sequencing the First Microbial Genomes

With all the recent news about genomics, some people are unaware that the first genome (*Haemophilus influenzae*) was sequenced only a few years ago. This was reported in a paper my colleagues and I published in *Science* in July 1995, and there is an interesting story behind it.

I spent almost a decade at the National Institute of Neurological Disorders and Stroke of the National Institutes of Health (NIH). NIH's intramural program is one of the best biomedical research programs in the United States. I had a multimillion-dollar budget and the freedom to work on anything I wanted to as long as I made discoveries related only to the human brain. The problem at the neurology institute was that we were discovering genes outside the central nervous system. This made senior officials nervous with regard to funding, which is allocated based on disease association.

In our work we developed the Expressed Sequence Tag (EST) method, which has changed gene discovery and is now considered the standard method for these explorations. But 10 years ago this approach was extremely controversial. James Watson made the statement that it is a technique that monkeys could do, and people did not like the EST method because it changed the pace of gene discovery. Whereas previously most of my colleagues and I might have spent 10 years trying to find *one* gene, EST accelerated the pace exponentially.

In 1992 I left NIH because I was given a $70 million dollar grant to form the Institute for Genomic Research, now known as TIGR. At that time, the big breakthrough in genomics was in mathematics. Our team at TIGR had developed a new algorithm to assemble large numbers of sequences. When we first started sequencing genomic clones, the biggest limitations were the mathematical tools for putting large numbers of sequences together and the small capacity of the computers available then. In the early 1990s it was difficult to assemble more than 1,000 sequences, and we had hundreds of thousands of EST sequences to assemble. Moreover, we knew that there were not that many genes; therefore, there must be multiple sequences per gene. We had to develop a new algorithm to put those sequences together and new computer programs to track the information. What we realized as a result of this effort was that we had created a powerful new tool that would allow us to go back and rethink genomics.

In 1994, Nobel laureate Hamilton O. Smith and I wrote a grant proposal describing a new approach for sequencing genomes and submitted it to NIH. We thought we could sequence the *Haemophilus influenzae* genome in one year. Keep in mind that back then the effort to sequence the *Escherichia coli* genome was in its ninth year of funding and that it took 12 years altogether to sequence the whole genome. Yeast took over 10 years to sequence with a major international effort. Needless to say, people were somewhat skeptical about this new approach. Hamilton Smith and I decided that we probably would not get funding, so we used money from the TIGR endowment to do the experiment. We had almost completed it when we finally got our review from NIH, saying that it would never work and would not be considered for funding. A month later we published the first genome in *Science*. One might think the story ended here but it did not, and it is important to understand some of the thinking and history behind advances in this field.

Challenging Preconceived Notions

We recently published the genome of *Vibrio cholerae* in the journal *Nature*. Aside from the technology used, almost every preconceived notion that scientists have had about every genome from any species was shown to be wrong as a result of this work. Some argued that sequencing *Vibrio cholerae* was a total waste of time and money because they thought 16S-rRNA and the cholera S-rRNA were the same as *E. coli* for all practical purposes and that there was one large chromosome that resembled *E. coli*. Therefore, scientists in the cholera field claimed that nothing would be learned from sequencing the cholera genome. But we sequenced the *Vibrio cholerae* genome anyway with funding from the National Institute of Allergy and Infectious Diseases because the institute believed that the whole-genome shotgun technology worked well with pathogens.

It turned out that the cholera genome had two chromosomes, rather than one. One chromosome closely resembles *E. coli*, but the other looks nothing whatsoever like *E. coli*. It probably carries most of the genes responsible for cholera being an infectious agent and being able to go into a dormant state.

A list of genomes that have been sequenced and published or will soon be published by our teams at TIGR and Celera includes a broad array of pathogens, some of which are environmental organisms, which from our human-centric view of life appear to have characteristics more akin to science fiction. For example, *Methanococcus jannaschii* is totally frozen at human body temperature. It comes to life at about 60°C, and the optimum temperature for its growth is

85°C degrees—it is completely viable in boiling water. It is a true autotroph and uses only two sources for its metabolism: carbon dioxide as its source for carbon and hydrogen as an energy source. This was found in one of the hyperthermal "black smokers" that are one and a half miles deep in the Pacific Ocean.

The other environmental organism we have studied—and a favorite of mine—is *Deinococcus radiodurans*. *Deinococcus* can take 3 million rads of radiation and not die; it is totally stable in a vacuum over years, maybe thousands or millions of years; and it is completely desiccant resistant. It can be totally dehydrated and can take huge doses of ionizing radiation in the dehydrated state. At first the chromosome gets blown apart with 100 or 200 double-strand breaks. Then, if dropped in an aqueous environment, in over 12 to 20 hours it stitches its chromosomes back together and starts replicating again. We would not have assumed this could possibly occur from a human biology view of life.

Francis Crick was one of the early proponents for the panspermia hypothesis—that is, that life actually originated somewhere else and came to earth. *Deinococcus* is a great candidate for that theory, so don't get too excited when NASA (National Aeronautics and Space Administration) announces that it has discovered *Deinococcus* on Mars or in outer space. Every time a traveler goes up in space or the commode on the space station is flushed, billions of copies of *Deinococcus* get launched into outer space. NASA just recently decided to do an experiment in which it will paint the outside of one of the shuttles with *Deinococcus* to see if it will survive in outer space.

We have barely scratched the surface of biodiversity in the extremes. We know of organisms whose genomes are now being deciphered and whose optimum temperature for growth is 1°C. *Thermotoga* is another example. It is a hyperthermophilic organism, whose optimum temperature is about 80°C, and which breaks down plant debris. It has a cellulose-metabolizing system. We found a large portion of its genome that was in the general prokaryotic category. However, a large portion of its genome came from the *Archaea* through lateral gene transfer. If one thinks of the evolutionary tree, particularly in the microbial world, there is always parent-offspring transmission of genetic information. If a lot of genetic information is moving around laterally between species, evolutionary trees are very imprecise models. Once we identify genes and genomes, we can find lateral gene transfer and determine how extensive it is in the microbial world. The concern is that it might be very extensive in the plant world, which has a lot of implications in terms of biotechnology of plants. TIGR sequenced one of the two first plant chromosomes to be sequenced, which was published in *Nature* in 1999. Early in 2000 we published the first insect genome—the first genome from a system with a central nervous system.

The second genome we chose to sequence, *Mycoplasma genitalium,* a human pathogen, was chosen for a specific reason. It appears to have the smallest genome of a free-living organism. We needed three months to sequence it. It has fewer than 500 genes, 475 protein-coding genes, and a number of RNA genes. A year later in Germany a second *Mycoplasma* was sequenced, *Mycoplasma pneumo-niae.* It was found that all the genes of *Mycoplasma genitalium* had a counterpart in *Mycoplasma pneumoniae,* but *pneumoniae* had 200 extra genes.

We all think that evolution involves adding on genetic information and complexity, but what we are finding is that most human pathogens probably started from a much more complex organism and eliminated genetic material during evolution. One test of this theory was to see if we could knock out the 200 extra genes of *Mycoplasma pneumoniae* and still have a living organism. We also asked a simple question: Even with *Mycoplasma genitalium,* are all those genes necessary for life? We naively thought we could come up with a molecular definition of life based on a minimal gene set. So we used a technique that is relatively simple to employ once you have the complete genetic code, which is called "whole-genome transposon mutagenesis." This technique uses electricity to incorporate transposons into a cell after which one can look to see where they incorporate in the genome. Some genes have a very large number of transposons; other genes have none at all. If the gene has none, we assume that it was probably essential for life. We assume that if a gene has some transposons in the middle of the sequence or a lot of transposons, it is dispensable. To make a very long story short, we got down to roughly 300 genes. The 200 extra genes in *pneumoniae* were completely dispensable, and about 200 genes in *genitalium* appeared to be dispensable.

Lessons Learned

This sequencing work yielded three stunning findings. First, out of the 300 genes we identified, 103 were completely new to science. We think we know a lot of biology, yet here is the most minimal cell and when we get down to it, we find that we have no idea what one-third of the genes do, except that if we knock them out, the cell dies. This was a very humbling experience for us. In the early 1970s when I was at the University of California in San Diego, I was told that it was going to be very difficult to find something new in biology because it was essentially all known. There have been other great pronouncements in biology. For example, in the 1970s the U.S. Surgeon General announced that we had won the war on microbes. We should never assume we know everything there is to know. Our work has demonstrated that one could randomly pick any one of these genes and create a lifetime career out of trying to study its function. Unfortunately, our research funding system discourages this kind of open-ended exploration.

The second thing we learned from this was that we could not come up with a molecular definition of life. It may sound trite, but we found that life is context sensitive; in other words, the environment that a cell is in is equally important to any components of the genetic code. A very simple example is *Mycoplasma*, which lives on glucose or fructose. If you knock out the glucose transporter gene and still have both sugars in the environment, the cell will live. But if there is only glucose in the environment and you knock out the glucose transporter gene, the cell dies. For each species, for each set of genes, there is a very precise set of environmental conditions or a broad range of them, depending on the specificity. So when we are studying the genetic code we are only studying at best half the equation. We humans have 100 trillion cells and maybe 50,000 genes, at least half of which are unknown and work together in these mass complexes. So it will be quite a while before we even understand the basic functions, let alone how all of them interact effectively with the environment.

The third thing we have learned is that Darwinian evolution is not just random errors in the genetic code. With *Haemophilus* and every pathogen we have worked on since, we found there was pre-programming in the genetic code to cause change in the structure of specific genes. That is why the Surgeon General was wrong. We have not won the war against microbes. We had a temporary gain that now has been almost lost. When we had the complete genetic code of *Haemophilus*, we looked at the genome and found tetrameric repeats in front of all the genes associated with lipopolysaccharide biosynthesis and on almost every cell surface antigen. Everybody has *Haemophilus influenzae* in his or her airways because it evolves in real time, due to the following mechanism: for every 10,000 replications the DNA slips on these tetrameric repeat regions. It changes the structure of the gene downstream, basically knocking it out. It fools our immune system by changing the structure of the lipopolysaccharides and it totally changes the antigens on the cell surface. Therefore, it is constantly evolving and fooling our immune system. All the pathogens have different types of mechanisms for doing this. These have real implications. After we sequenced *Haemophilus*, a company tried to make a new vaccine against the microbe but ignored these findings. As a result, a great vaccine against the parent strain was developed that failed as soon as it went into clinical testing because we each have a slightly different strain variant due to these mechanisms.

In collaboration with Chiron, TIGR has been developing a new vaccine for meningitis, based on the sequence of the *Meningococcus* genome. At the same time we published the complete genome we also published a study on the vaccines. Two vaccines were developed in less than a year by using two cell surface label proteins that did not have these variation mechanisms. Thus far these vaccines seem to be effective against all the different strains in clinical trials.

The Whole-Genome Shotgun Sequencing Strategy

Gradually, it became clear that we had developed a robust technique based on mathematics and computing and a little bit on new molecular biology. We were looking for a way to scale up when Applied Biosystems called me in the fall of 1998. The company had developed a new DNA sequencer and was willing to give me $300 million to do the experiment I wanted to do to sequence the human genome. The catch was that we would not be funded at our not-for-profit institute; we had to form a new company to do this. The whole-genome shotgun sequencing strategy allowed us to form Celera Genomics and to sequence several genomes. But probably equally important is the advent of high-end, 64-bit computing. New algorithm development is probably the most important key to going forward in this field. We have now hired 40 of the top algorithm scientists in the world.

We set up a large sequencing factory. It took about six months to build our facility and totally equip it. Our laboratory is the size of a football field and full of machines. Basically there are three components: the DNA sequencing machines, massive amounts of electricity and air conditioning, and a fiber connection to our computer facility. We bought 300 machines at $300,000 each. In contrast to the more than 1,000 scientists in the public effort, we initially started with 50 scientists who ran all the sequencers. Now we are down to 9 scientists who do 200,000 plasmid sequences 24 hours a day, 7 days a week. We have substituted electrons for people and initiated very new high-throughput efforts.

As a biologist I didn't know anything about high-end computing. But fortunately I am an experimental scientist, because I had to evaluate all the major computer manufacturers in the world to try and work out which computer might be able to assemble the human genome. There was no way to sort out the claims from IBM, Digital, Sun, and Silicon Graphics, so I gave them a problem to solve. I gave them the *Haemophilus* genome and our algorithm and asked them to see if they could improve on the 11 days it took us to assemble it with a Sun 32-bit computer. Only two computer companies, IBM and Digital, could even run the experiment. IBM's best effort got it down from 11 days to 36 hours. With some optimization with the alpha chip, Digital got it down to 9 hours. Eleven days to 9 hours was a big improvement. So we worked with Compaq to build a massive facility. We now have over 1,200 alpha processors. The current database size is about 80 terabytes of data (therefore, any dreams people had about getting their genome on a CD-ROM won't happen). In addition, we bought a parallel computer company that makes custom computer chips. It designs our custom processors and optimizes them for sequence comparisons. And the company built a second unit that optimizes text searches. We need this million parallel processor computer to daily download the world's literature to annotate and update the human, mouse, and other genetic codes as they are developed.

Going straight from the microbial to the human genome was a big step because when we set this up we didn't know for sure that the DNA sequencers would work. We had only seen an engineering prototype. But I had confidence in the capabilities of the engineers and was sure that they would work eventually. The mathematics were a big challenge. We couldn't use the algorithm we had developed at TIGR because it wouldn't work at the scale we needed. So we developed a whole new algorithm for putting the genome together. We decided to try it with the *Drosophila* genome, which was the largest genome being studied.

Use of the whole-genome shotgun technique was remarkably simple compared to the way in which the public effort was proceeding. We basically take all the DNA out of the cell, use mechanical shearing to cut it into different-sized fragments, select fragments by size, and ligate them into plasmid vectors. All our sequencing is double stranded—that is, we sequence from both ends of each clone. This is critical to how the process works mathematically because we use different-sized clones, ranging from 2,000 to 50,000 letters long. We developed a cloning technique for 50-kilobase clones. Doing tens of millions of sequences meant we had to have absolutely foolproof software tracking to make sure we could keep all these ends associated with each other. We also used backend sequences.

While the rest of the community was fretting over the repeats in the human genome, we realized that if we ignored the repeats we could unambiguously assemble at least 99.7 percent of the genome and then come back and deal with the repeats. The technique is remarkably simple mathematically. It is just doing the linkage comparisons: if the end of one sequence overlaps with the end of another sequence, you build structures of different sizes. For the *Drosophila* genome we had 3 million sequences and for the human genome 45 million sequences, and we only put things together where there was a single mathematical solution. We have sequences that are 500 to 600 letters long. Imagine trying to do this by hand! Imagine trying to line up 45 million of those sequences in terms of working out where the overlaps are, especially when there are a lot of repeats in the genome. So we only put things together where there was a single mathematical solution in the entire human genome. There was less than one chance in 10^{15} of making an error. That is why we were so confident that this would work even when everybody else was saying it was absolutely impossible.

We ended up with scaffolds built from these approaches that only had small holes where the gaps were. One of the reasons we chose the *Drosophila* genome was because it was the best mapped genome. Therefore all these markers were mapped very accurately on the genome. We did a comparison of all the markers and found that only 16 did not agree with our sequence assembly. After analysis, the mapping community in *Drosophila* went back and found that every one of these errors occurred as a result of the mapping method; not one was an error in the sequence assembly.

Having all these data in hand in a very short period of time created new demand for how we were going to annotate and interpret it. We convened what we called an "Annotation Jamboree," where we brought top *Drosophila* scientists from around the world to Celera. All of these scientists were experts on different gene families or different specialties in the genome. Daily and nightly we went through the genetic code, and in less than a year the *Drosophila* genome was published. The next largest genome to be completed was the *C. elegans* genome, which took over a decade to do in the clone-by-clone approach.

Some numbers on this: with *Haemophilus* we had to sequence 26,000 sequences. That was a big deal in 1995; it took four months. We sequenced the 3 million *Drosophila* clones in four months. If we repeated this experiment today, it would take roughly three and a half weeks to sequence the *Drosophila* genome. If we were going to resequence the *Haemophilus* genome today, it would take two hours, and if we were going to redo the yeast genome that took 10 years, we could do it in eight hours.

There were roughly 2,500 genes known after a century of research on *Drosophila*. By the end of the Annotation Jamboree we had characterized more than 13,000 genes. A lot of those are of unknown function. On average, 47 percent of the genes on each genome are new to science. There are no homologues; they don't look like anything we have seen anywhere before, except roughly half of those match other unknown genes. When we looked at the *Drosophila* genome, only 40 percent of the genes could be identified by any clear-cut homologous or paralogous searches. That is, only for about 40 percent of the genes can we come up with any reasonable description of their biology. Sixty percent of the genes are totally new. We have no idea what they do. Yet they are responsible for a multicellular organism with a complex central nervous system, and the human genome looks remarkably similar to this.

Functional Genomics

How are we going to go through the tens of thousands of genes in each of these species? How are we going to learn their function? How are we going to learn their biology? Out of the work done on microbial genomes there have been more than 100,000 new genes introduced that are of unknown function. In the year 2000 the human, mouse, and *Drosophila* genome systems introduced tens of thousands more. This is a challenge for the pharmaceutical industry. A lot of these genes cause disease, even though we have no clue about their function. NIH spends about $2 billion a year funding single-gene cloning projects like the kind I spent 10 years of my life on, which they no longer need to do. What took me 10 years I can now do with a 15-second computer search. We are starting from a different point in science. We have all this basic information that can no longer be ignored.

In the 1970s everything in physiology and medicine was explained by cyclic AMP levels going up or down in cells. Now we have slightly more complex information to deal with. It is a real challenge for all of our major public and private funding agencies to find ways to understand this information. Even on the so-called known side of the genome, we do not truly understand function.

As an example, Seymour Benzer characterized one gene in *Drosophila* that he named Methuselah that greatly increased the life span of fruit flies. When we sequenced the *Drosophila* genome, we found 11 Methuselah-like homologues. Everybody in the Annotation Jamboree who was over 50 years old switched immediately to the human genome to see if we could find Methuselah-like homologues.

Moreover, we found over 300 human disease genes that have their best counterparts in the fruit fly genome. There were roughly 6,000 *Drosophila* scientists before we published the *Drosophila* genome. If we count the hits on our database and all the other public databases where we have posted the sequence, there are about 100,000 hits every day. When we published the *Haemophilus* genome, there were only two research laboratories in the United States studying *Haemophilus*, even though it is a key pathogen (it causes ear infections and meningitis in children). There are now thousands of labs around the world studying *Haemophilus* because the genetic code is available. This information will categorically shift science forever. The question is: How long will it really take our funding and science systems to adapt to the information available?

Sequencing the Human Genome

We sequenced the first human individual three times because we were worried that genetic variation would affect the mathematics of the assembly. *Drosophila*, yeast, *C. elegans,* and every other species that had been sequenced before were highly inbred strains. There are no variations in the chromosomes. But with each human individual, each chromosome differs from that of other individuals in roughly 1 in 1,000 letters. If we randomly sequenced 10 chromosomes of numerous people, we would introduce a very high variation rate, which complicates the mathematics. We sequenced the genomes from three females and two males. With these paired clone coverages, we covered the genome about 45 times.

On June 26, 2000, the White House event to announce the first rough draft of the human genome took place. In conjunction with the public effort, represented by Francis Collins, Celera announced that we had assembled 3.12 billion letters of the human genetic code using our computer system and the new algorithms. This was an important event because it quelled the bickering in the scientific community, especially from people who did not want the techniques to change because they were worried it would affect their government funding or because they really didn't think this approach would work. There has been good cooperation ever since, and we simultaneously submitted both genome efforts to a scientific journal in late 2000.

Studying Multiple Organisms for Similarities and Differences

In the 1990s we found three new genes that cause colon cancer. These genes were identified by comparing the human sequences to those from yeast and *E. coli*. Because common mechanisms in our cells have been highly conserved over 3 billion to 4 billion years, the sequence homology in these DNA repair enzymes was extremely high between human and bacteria. A lot of people question the value of funding research on yeast or bacteria. But by studying the biochemistry in simple systems we were able to find these genes associated with colon cancer. In fact, we will develop most of the knowledge from the human genome by comparing the sequence data with those from other species. For example, the gene order on the human X chromosome, the mouse X chromosome, and the cat X chromosome are nearly 100 percent identical. In fact, with over 90 percent of the mouse genome sequenced, we have not yet found a human gene that does not have a counterpart in the mouse genome. It is not clear how many human-specific genes there will be. People are very anxious for us to sequence the chimpanzee genome to see what happened during primate evolution. The indications are that it is not changes in proteins that differentiate the species but rather changes in the regulatory regions. While you would expect the 3 percent of the genome that codes for proteins to be the only thing that is really conserved between mouse and human, we find regions where 20 percent of the genomes are nearly identical: the letters, the genetic code—key pieces in terms of the structure, function, and regulation of human genes. We are using this information to accurately determine the number and structure of genes in the human.

As another example of the interconnectedness of genomes, if you knock out the *pax-6* gene in *Drosophila*, it leads to what is called an eyeless phenotype. These fruit flies are blind. If you knock out or mutate the same gene in mice, it leads to blindness, and if the same gene is mutated in humans, it leads to a disease called aniridia, in which babies are born without an iris; thus, they cannot regulate the light going into their eyes. You can take the intact human or mouse gene and put it in the fruit flies and it rescues the phenotype. Not only is the sequence similar but the proteins produced have very similar functions. That is why tools like fruit flies are so important in terms of studying human disease.

Human Variation and Drug Responses

With the human genetic code we find roughly 2 million to 3 million variations in the chromosomes. We have about 2.8 million well-characterized so-called SNPs (single nucleotide polymorphisms) in our database that are now being used by scientists around the world to study linkage to disease. For the first time we can look at this genetic variation by chromosome. For example, you can discover genetic variation in the genome of individuals that have an increased risk for myocardial infarction. The pharmaceutical industry is extremely interested in using this information to find ways to improve clinical trials and drug effects. This could lead to what we call personalized medicine, or pharmacogenetics.

As another example, a type II diabetes drug recently had to be taken off the American market because 1 out of 10,000 people had a severe liver toxicity to it. If we can find simple tests that predict toxicity it will have a huge impact. Not only can we change adverse drug effects in the population, we can also tailor drugs so that they work for more than 30 to 50 percent of the target population, the current average. Currently, we administer drugs to all candidates based on the assumption that they are relatively safe. Personalized medicine offers the potential to give individuals drugs that will actually treat their disease and not cause serious side effects or even death. The leading pharmaceutical companies recognize that this is the correct ethical and scientific approach that will lead to increased diversity of drugs.

Where Are We Going in the Future?

I want to put these genetic changes into perspective because there is a lot of hype about the genome project. With the DNA mismatch repair enzymes that lead to nonpolyposis colon cancer, we can measure those genetic changes in the population and can tell you whether you have an increased risk of getting colon cancer. We cannot tell whether or not you will get colon cancer. That is going to be the big challenge in terms of applying these tests broadly to the field of medicine because we are going to be dealing with probabilities and risks and not absolutes. The only way to get to absolutes and narrow down the cost factor is to use genetic screens to find out which part of the population has increased risk and then do the more expensive tests on those individuals. For example, with colon cancer, if you learn at age 20 that you have an increased risk, you could get a colonoscopy every year. One colonoscopy costs about

$1,500 and unless you have symptoms, most insurance companies will not pay for it. Thus, having more narrow and certain estimates of risk will improve health and save money.

The challenge is to come up with early markers for disease. This is the field of proteomics. It is an old field that used to be called protein chemistry, yet it is also a new field because now that we have the genome we can get at every protein. If there are 50,000 genes there are somewhere between 200,000 and 1 million proteins. We are building a large-scale protein facility to do roughly 1 million protein sequences a day. The reason the genome is important is that with mass spectroscopy sequencing the proteins get blown apart into small fragments, and we can compare those sequences with the databases. Until now most of these did not match anything in the database, so we could not interpret the data. Now every one of these will have a match, and we can rapidly determine the sequence of the proteins in the cells and the blood. Current technology can do 100 samples per hour. New machines will be able to handle 10,000 samples per hour. We are preparing to process over 1 million protein sequences per day from as many as 10,000 patients.

These efforts are all multidisciplinary. We employ more computer engineers and programmers than biologists. We have more engineers and physicists designing and building new machines than technicians to actually run them. The limitation of all these is going to be mathematics. The challenge with genomics is small-scale compared to deconvoluting all the data that are coming out of these machines. But we are confident that this will lead to new treatments for cancer on a personalized basis, to cancer vaccines against very specific sequences in proteins. In 1998 we started with genomes. In 2000 we sequenced the *Drosophila* genome and published that, and we also sequenced the human genome. We are scaling up now to sequence proteins on an even larger scale. The real breakthroughs that have allowed all this have come from mathematics, computing, and physics. We are hoping these efforts will lead to dramatic changes in the world of medicine and human health.

Leroy Hood

After the Genome

Where Should We Go?

This is one of the most exciting times in biology. The revolutions that have been generated by the first draft of the Human Genome Project have barely been felt, but there is one profound change that has already occurred, and that is the realization that biology is fundamentally an information science. I would argue that this realization is key to understanding where we are going to go in the future with the Human Genome Project.

Almost 40 years ago Ed Lewis discovered a remarkable fly that differs from an ordinary fly by one extra pair of wings. It turns out that the mutation that causes this extra set to grow is a single mutation in a single gene. What that says unequivocally is that biological information is hierarchical. It is hierarchical in the sense that some units of information can affect extreme, complex kinds of changes.

But fascinatingly enough, in this particular case it also says that biological information is historical, because the fly with the extra wings is the evolutionary antecedent to the contemporary fly. Understanding the hierarchical nature of biological information and applying it to science and medicine is what will change our future.

Origins of the Genome Project

I was at the first meeting held on the Human Genome Project in the spring of 1985. Robert Sinsheimer, chancellor of the University of California at Santa Cruz, had raised $35 million and was considering the formation of an institute dedicated to sequencing the human genome. He invited a small number of scientists—Walter Gilbert, George Church, Charles Cantor, and David Botstein—to meet and consider the topic and its various ramifications.

I was slightly skeptical on technical grounds but came away with two deep convictions. First, this effort had the potential to transform both biology and medicine, and it would drive an enormous technology development effort. We needed to invent new and better machines and computers for analyses. Second, I saw in the Human Genome Project the introduction of a new type of science in biology—that is, "discovery-driven science." Discovery-driven science, as compared to hypothesis-driven science, takes an object and enumerates its elements irrespective of any questions. That is, it creates an infrastructure on which hypothesis-driven science can be done far more effectively.

There was enormous acrimony in the first five years when a few of us pushed the Human Genome Project, and I am convinced that a lot of this acrimony and misunderstanding was centered on misunderstandings about the power of discovery-driven science. Those misunderstandings are still reflected and embedded very deeply in the cultures of national funding agencies. What the Human Genome Project offers us is what I call "systems biology," that is, integrating hypothesis-driven and discovery-driven science.

Nature and Nurture

The human genome is the world's most incredible software program, a program in which a single fertilized egg can create an adult organism with 10^{13} or 10^{14} cells. The genome is played out through a type of "chromosomal choreography," in which different cells and subsets of the genes are expressed, thereby manifesting their distinct phenotypic potential. This is the arena of developmental biology, and of course it is going to be fundamentally altered by the Human Genome Project.

The really critical question raised by these emerging views is: "To what extent are we our genes?" I remember James Watson about 10 years ago making a statement: "We used to think our fate resided in the stars. We know now it resides in our genes." But the importance of nature and nurture is fundamental. Its ambiguity is illustrated in the example of fingerprints of identical twins, which are quite different from one another despite the fact that the genes in the two individuals are absolutely identical. What this says in an unequivocal manner is that when the developmental program for fingerprints unfolds, either as the consequence of stochastic events or different environmental signals, quite different outcomes occur. The tools for understanding the relative roles of nature and nurture in the development of phenotypic traits are horribly inadequate at this point in time.

Signal Contributions of the Human Genome Project

I would argue that the genome project has made three major contributions already. One, already mentioned, is in the field of discovery science—it is fundamentally altering our view of how to do science and biology.

The second contribution is in the development of a "periodic table" of genetic elements, with four major types of information. The first is the 50,000 to 100,000 or so genes that are present in the human. Second is the sequence information that surrounds those genes, because embedded in those regions is the regulatory machinery or code that is critical for turning genes on and off. Third, we can take the genes and deconvolute them into their basic building block components, or "motifs." These motifs, of which there may be 1,000 to 2,000, constitute the "Tinker Toys" or the building block components for understanding how genes are assembled and even what gene functions are carried out. This information will help us understand one of the central problems in modern protein chemistry— that is, how proteins fold and how that configuration affects the structure and function of the building blocks. Fourth, we are gaining information about normal and abnormal human variability.

But in many ways what is most transforming are paradigm changes that have altered the face of biology. I have already mentioned that biology is an informational science, but let me emphasize again the hierarchical nature of this information. We progress from a gene to a messenger RNA, to a protein, to informational pathways that carry out specific functions, like the metabolism of a particular carbohydrate, to many such informational pathways that are interconnected in an informational network. It is the operation of this network that gives that cell its phenotype.

The information present at the gene level cannot necessarily predict all of the information present at higher levels. Proteins are modified, interact with other proteins, and are compartmentalized, so these aspects of their informational code cannot be predicted from the primary sequence of genes themselves. In addition, the informational pathways have so-called systems properties that again cannot be predicted from knowing the sum total nature of the individual units. Thus, the idea in systems biology is to understand not just the individual components in the system but rather their function in a system.

A second paradigm change is the concept of "high-throughput biology," which has evolved from the sequencer prototype that Lloyd Smith and I developed in 1986. That instrument had about 1/2,000 the throughput of the capillary sequencers used today.

Additional changes have occurred in improved data quality,

decreasing cost of obtaining data, and the emergence of cross-disciplinary collaborations in engineering, chemistry, computing, biology, and physics.

The Future of Sequencing

I can envision within the next five years DNA sequencers on little chips the size of your thumbnail. They will have considerably greater capacity than our contemporary instruments, and into the future we may have powerful new methodologies for reading out the letters along the fragments of single DNA molecules. In San Francisco there is a company that has already developed the capacity to simultaneously sequence 1 million sequences for 16 to 20 base pairs. In the near future we will be able to put all of the human genome on a single DNA chip. The power of the chip is that we can take a normal cell and a cancer cell and look at every single gene in the organism and ask: How has it changed quantitatively? Is it like the cancer cell or the normal cell? We can visualize the changes.

A second vision of the future involves the contribution of computing to biology and vice versa. Computing has become essential to contemporary biology. We need the tools of computer science to acquire, store, analyze, decipher, and graphically display DNA sequence models for distribution. What is fascinating is that living organisms have had about 3.7 billion years to manipulate their "digital strings," and in doing so they have invented digital strategies that are going to turn out to be incredibly useful to people in computer science. It is this attractiveness that has brought first-rate scientists from the realms of computer science and applied mathematics into biology, so it is a two-way contribution.

Evolution as a Tinker

The importance of animal models will continue to grow as they help us decipher complex biological phenomena. Indeed the gene that was mutated to convert a wild-type two-wing fly into a four-wing fly was a member of the family of genes called HOX, which are important in the regulation of the axial development of the fly, and it turns out, in the axial development of the mouse and the human as well. The really striking idea is that we can gain fundamental insights into how humans develop by studying how flies develop. The strategies are very similar, even if the outcomes are different in their external manifestations, as a consequence of different regulatory strategies.

We now have genomes from many different organisms and are in a unique position to begin in a much deeper sense to understand the logic of life. We have the ability to use computer tools to deconvolute chromosomal strings of model organisms into their individual genes and then to place, at least initially, those genes in their informational pathways to understand the nature of the logic of life in this particular organism. The ultimate goal of comparative genomics is to put side by side two different logics of life to understand how biological mechanisms operate and the nature of their constraints. This has to be superimposed on biological evolution because what we know from all of these genomic sequences is the incredible fundamental unity of life and the fact that at the very beginning the basic strategy and rules for simple biological mechanisms were laid down only to be elaborated in many different ways by more sophisticated kinds of organisms. As the scientist Max Delbruck once said: "Any living cell carries within it the experiences of a billion years of experimentation by its ancestors." We now have the ability to look forward to deciphering that history and coming to understand that particular biology.

We do not really understand the origins of biological information, but I suspect there is a lot of inadvertency in those early events and in the choice of which particular nucleic acids and amino acids were employed when life began. There were early constraints on the system that set the pathways for the development of the biological simplicity we see today—the As, Cs, Gs, and Ts. What is obvious is that the first informational molecule that really arose was not DNA or a protein, but RNA. RNA actually has two informational properties: first, it reads the digital signal that is the essence of DNA, and second, it can fold into three dimensions and catalyze its own synthesis.

Precursors available in that early prebiotic sea combined with events occurring in the clays in the thermal vents to catalyze the first primitive RNA-like molecules. Slowly and gradually, RNA molecules evolved to have the ability then to catalyze themselves. One question that no one can answer is how they evolved to package this information in a membrane so you could contain in a concentrated fashion the information, energy, and catalytic reagents.

Once the requisite elements for reproduction were assembled, those informational molecules had an enormous advantage over anything else that was out there, and so they became dominant. One of the interesting questions is: Were there alternate subunits and systems that lost out through natural selection? I think the essence of evolution is that it is a tinker, and there are numerous solutions to survival, but once a successful solution is reached, everything that happens subsequently is built on that successful solution.

Deciphering the Genome Using a Systems Approach

So where do we stand today with regard to the genome project? The first stage of sequencing a genome is reading out all of the different letters in the chromosome string. The second stage is then using limited biological approaches, and some computational approaches, from that single undifferentiated string to begin to fashion the words and even the punctuation that can put these words into sentences and possibly paragraphs. We are somewhere at the beginning of this stage right now, trying to make sense of strings of undecipherable information. Systems biology will help us to convert unintelligible information into knowledge about biology.

Using the analogy of understanding how a car works, we would approach it with a systems approach by breaking the car down into all of its individual elements and then perturbing the system. In the car we would have it carry out its functions—for example, go forward, go backwards, and brake. And then we would measure the relationships of these elements one to another, and in biological organisms those measurements are much more complex because of the hierarchical nature of biological information. Then we have to integrate this information and formulate an initial model that begins to predict the structure and behavior of the car. The model itself then suggests new kinds of perturbations that we use in a cyclical process to improve the model. Ultimately we can create a model that will do two things: define the structure of the car and define its systems' properties, given a particular kind of perturbation.

If we translate this simple analogy into approaching biological systems, the one critical point is that the iterative nature of this repetitious model-building cycle requires close juxtaposition of the physicists, computer scientists, and engineers with the biologists. This is one of the challenges of the new biology—how to breach the language barriers and bring cross-disciplinary colleagues into close juxtaposition. By doing so we can better understand biological pathways that have been studied, in some cases for more than 30 years.

In addition, understanding these pathways will lead us to elucidating the regulatory code of systems—that is, why organisms differ even though they are all made of DNA. For example, the human and the chimpanzee differ from each other by roughly 1 percent of the DNA sequence. The structural gene variations are minor, so the key variations have to be regulatory in nature. The systems property of regulation addresses such phenomena as when genes are expressed during the developmental stage of the human, in which cells in the organism they are expressed, the amplitude or magnitude of expression, and the ability to be expressed coordinately with many other genes in the networks and systems.

Using this type of knowledge we can, for example, apply it to the study of the immune system and its two major players, adaptive immunity and innate immunity. By triggering the basic cells of the immune system to carry out their various systems properties, we can then interrogate by systems biology the nature of the informational pathways and understand the changes that occur in immunity, tolerance, or autoimmunity. Although we probably know more about the molecular details of the immune system than any other system, because we have not taken a systems approach to its study we do not understand fundamental systems properties, which is why we have been unable to produce effective vaccines.

I believe there is going to be a revolution in our understanding of the brain with these new types of systems approaches. The brain presents particular problems because of its integrated coherence, and how to get out individual cells to interrogate for their properties is an enormous challenge. But the beginnings of new approaches to understanding the human brain are under way.

Preparing for Changes in Medicine

One of the most fascinating problems raised by the study of genomics is the question of human variability. What is the nature of the variation in our genes that leads to such different phenotypic consequences? We now have the ability to look at the variations that predispose us to disease, and as we do this over the next 25 years we will fundamentally change the nature of medicine from reactive, to predictive, to preventive.

In the beginning we will not have the preventive measures— that is, we will be in a predictive state where we can do the diagnostics and write out the probabilistic possibilities for future health histories, but until we develop the preventive measures, we will not be in a position to actually practice preventive medicine. The key for going from prediction to prevention is putting defective genes in the context of the informational pathways in which they operate and then, through an understanding of the pathways, generate the corresponding preventive measures.

What is quite clear is that this method is going to let people live longer. How is society going to deal with an aging population that may be very creative and potentially contributory? How are we going to change the training of physicians? One of my favorite questions is to ask a physician audience "to write out your job description in the year 2020," and I can guarantee you it will be transformed totally from what physicians are doing today.

How do we educate society? Society sets the constraints on where science goes and the resources with which it can move forward, and of course society shapes the ethical, legal, and social conversations. One of the most important societal issues will be educating our children in science so that they can integrate this information into all aspects of their lives.

We stand at a transforming point in the history of biology. In the 1970s, Gordon Moore made the prediction that the number of transistors that could be put on a computer chip would double every 18 months, and that prediction has been true for the past 30 years. This accomplishment more than anything else has driven the revolution in information technology and communication. We are now in a similar position with regard to biology, where we see an even sharper exponential increase in the amount of DNA sequence information to come. The challenge is in how to convert this DNA sequence information into knowledge. The key is to be able to acquire knowledge about systems from all the different hierarchical levels to come to an understanding of the nature of the systems that make us uniquely human.

The other point I would make is that the hierarchical nature of biological information extends beyond cells and organs and even individuals in populations and ecologies. I would argue that systems biology offers a wonderfully integrative view of how we can bring all of the levels of biological information together in a uniquely powerful fashion.

Part II

Applications of Genomics to Medicine and Agriculture

Robert Bazell

Introduction

North of New York City, in Buffalo, there is a well-established institution called the Roswell Park Cancer Institute. Many people hear the name and think there is a park called Roswell Park, much like Central Park. In fact, a surgeon named Roswell Park founded the institution. He was able to garner support to build the center by going before the New York State Legislature and saying: "The cure for cancer is just around the corner."

We should be suspicious of anyone who makes that claim today because it has been said so many times by so many people. I am not going to tell you that the cure for cancer is around the corner because it is not. However, better treatments for cancer are so close that it is causing an excitement unlike anything that has happened in medicine in a long time. In this section, William Haseltine and

Arnold Levine tell the story of how the new genomics has evolved to bring us the promise of new therapies for cancer and a host of other diseases.

By combining evolutionary history, family histories, and medical history, geneticists such as Mary Jeanne Kreek are able to uncover the genetics of complex diseases like addiction. Kreek relies on family histories as well as the wide array of single nucleotide polymorphisms (SNPs) now available to geneticists to demonstrate how a complex disease of the brain can drastically affect human behavior, which can be manifest as addiction. Complex diseases like cancer or addiction are the result of the coming together of many variants in the genome at many different genes.

Finally, the use of genomics in agricultural biotechnology has already yielded improvements in pest resistance and drought resistance and can be used to preserve biodiversity. Barbara Schaal explores the application of genomics to agriculture, where there is tremendous potential to feed the world.

The scientists presented in this section have made enormous contributions to understanding the molecular bases of human disease and plant biology. And while we applaud the great promise of genetics to prevent, detect, and treat cancer and to feed the world, it is inevitable that such promise comes with possible dangers. Our past informs us that there is a dark side, that potentially harmful things can be done with genetic information. While we applaud the accomplishments, we must also proceed in a manner that ensures genomic information is put to good use. If we do, we can actually start to believe that better health is just around the corner.

William Haseltine

Genomics

Rapid Road from Gene to Patient

Never before has there been as exciting a time as now in terms of understanding nature and using that understanding to improve human health and other aspects of human existence. It is not just genomics that is racing ahead. Progress is also being made in other aspects of our technical mastery of the world—understanding the atom and understanding the chemistry that will lead to self-assembling, atomic-scale structures, which we can use for communication, computation, and eventually biology.

In biology we can use these tools to do what life itself does: assemble itself out of atoms that are precisely placed in three dimensions. Communication technology will be used not only to communicate between people but also to acquire large amounts of information about genes and biology and to sort through those data in a rapid and determinative way.

We who are in this revolution believe we are at the beginning of a golden age that will eclipse the past, as glorious as we may have thought it to have been. This is a great moment for humanity and a great time for science to share its excitement and communicate its knowledge. Those who will be the beneficiaries and the users of these advances must understand the powerful prospects for good that this technology brings, or it may not come to be.

Genetic Variation and Nature Versus Nurture

The word "genome" has in it the word "gene." A gene is often understood to define a difference. We cannot escape the realization that we are different from one another. We have come to learn that many of our differences—the shape of our face, the size of our body, how we age—are determined by our genes, which are passed on from generation to generation. We have also come to understand that the differences we see on the surface are reflective of far deeper differences in our bodies, differences that do not affect merely our appearance but also other aspects of our life and health. These include differences in the probability of developing cancer, bone disease, or psychiatric diseases; having a long life span; and becoming obese. In fact, the realization that so many differences exist between us is beginning to be daunting. The catalog seems to be continually lengthening.

These differences also highlight the age-old question of nature versus nurture. How much of what we are—whether we like corn flakes or oat bran in the morning, for instance—is determined by our genes? There may be genes that help determine that; it is not an outlandish proposition. We are now, through the powerful new tools that have been developed, on the brink of being able to associate any measurable difference in human behavior, phenotype, or disease predilection with some spot among our genes.

This is beginning to have a philosophical impact on our definitions of free will and what we are. At the same time, we are beginning to realize that although genes create potential, it is experience that creates our bodies. In particular, the organ we thought was the most static—the brain—we now know is one of the most dynamic. There is a potential created in our brain by the genes, but it is experience that creates the brain's physical reality. So there is a duality in our understanding of ourselves as products of our genes.

We cannot ignore the differences between humans and other species, and among those species, because the differences are many and great. One of the tasks that lie before us is to go beyond physical observable differences that can be measured with a caliper or protractor to understanding the biochemical differences and the gene differences. Then we will be able to redraw the tree of life. New insights mean we are already beginning to accumulate a much more complex view of the relationships among organisms. Species did not all stem from a single source and branch out: there were multiple crossing points in the history of life.

The common view of genes as determinants of difference must therefore yield to a far more powerful concept of the gene—the determinant of our common human heritage. Genes produce not merely the differences that separate us but also the deep underlying structures that unite us as a single species and unite us with all other species. From the point of view of DNA, there is only one life on this planet. It is whole and unitary. Even the distinction between animals and plants is secondary. The goal of much of science is to focus on differences, but our scientific insights can lead us to see that genes also provide unity.

The Gene as an Anatomic Object

The Human Genome Project will help us to understand and to quantify our differences, to predict what diseases we may get, and to level the genetic playing field through genetic therapy and gene manipulation. There is another more powerful and immediate consequence of understanding genes, which goes beyond the concept of the gene as a determinant of difference. This view sees the gene as an instruction to make a small part of human anatomy. It is an extension of the view of the anatomist that says life in any form is a wonderful working machine. This approach will tell us what the parts of the machine are, so we can take them apart, dissect them, and understand how they interrelate.

We can then use that knowledge, either to improve the plant or to cure the human. This is a profound revolution in our understanding. We are looking not at one gene at a time but at all of our genes, not at genes as inherited objects but as anatomic objects. This revolution will allow us to find genes that determine our structure and our function and that control our anatomy and our physiology. Sometimes these genes go awry, not because they are inherited but because the slings and arrows of outrageous fortune have altered them.

We know there are genes in our bodies that cooperate with microorganisms to cause illnesses we know as viral, bacterial, and parasitic diseases. These are not simple invasions of our bodies. They are two sets of genes working with one another in a cooperative fashion.

We now have for the first time a comprehensive view of these structures, which inform our body, build it from a single fertilized egg, maintain it, and repair it as we age. This view comes not as a result of sequencing the human genome; for various technical reasons, that turns out to be a poor way to understand genes in their totality and as they are used. But because we can now capture the edited form of genes in our tissues and catalog where they are used and under what circumstances, we can now learn how they change in health and disease.

This approach involves an interpretation of the genome that is different from that of the Human Genome Project. This approach sees the genome not as a complete inherited text with variations but as a collection of genes as they do their work in the human body. Through the power of modern molecular biology we have the opportunity to take each gene as it is used in the human body and use it to make an unlimited amount of the unique substance that it gives rise to in the body.

A Natural and Rational Approach to Medicine

A gene is an instruction that makes a protein. A protein is what does the work in any living system. If you look at your hand or your face, you are looking at protein or the products of proteins. We are made of about 120,000 different proteins working together, and each gene makes one of those proteins. We now have in our freezers at Human Genome Sciences, for the first time, a copy of almost all of those genes in a form in which they can be used. We know where the corresponding proteins are made in the body, and we have some appreciation of how they change as a result of changes in gene activity. That knowledge of similarity can be used for medicine.

For example, when most people think of insulin, they do not think of a human part in a bottle—of a manufactured human component. But that is how we in industry think of it. Insulin is a human protein made by a gene. Modern technology has allowed us to slice that gene out of one particular individual and implant it in a separate organism. Because of the unity of all living things, we can put a gene from a human into something that is removed from our own bodies by 2 billion years of evolutionary time—a bacterium or a yeast—and that organism will make the corresponding protein, be it insulin or something else. Thus, the unity of life can have very practical consequences.

Just as remarkable, insulin made from a single gene from one individual can be used to treat all human beings. That is indeed unity. At the level of our genes, we are much more similar than we are different, and we can use that principle as a powerful tool in medicine. It is our similarities, not our differences, that are our brightest hopes for the future.

Insulin is not an isolated case. We know that growth hormone, made from one person's gene by methods similar to those used to produce insulin, can be bottled to treat many people. The same can be said for erythropoietin. We have also learned to manufacture human antibodies in test tubes to treat infection. Moreover, we can modify the course of an illness not only by supplying something that is missing but also by antagonizing something that is made when it should not be made. A good example is the use of a human antibody, herceptin, to treat breast cancer.

Let me give you an example from our own work of how this new systematic knowledge can rapidly advance medical science. With medicine you always start with an unsolved medical problem. In this case the problem we started with is the desire to increase a human being's ability to fight infections. Our thoughts were focused particularly on older people, who lose the ability to respond to new infections by mounting effective antibody responses. There are many others who also could benefit from an increased ability to produce antibodies. They include AIDS patients, people recovering from chemotherapy (which damages the immune system), people with inherited defects in their ability to make antibodies, and many people who are fighting antibiotic-resistant infections.

The body must make a substance that causes the immune system to produce more antibodies. Try as they might, scientists in hundreds of laboratories around the world failed to discover that substance by classical approaches. Our approach started with our nearly complete collection of human genes in their useable form. We then selected a subset of those we believed had characteristics of signals that cause cells to behave differently—a subset of about 10,000 of our 120,000 genes. We isolated and made small amounts of the protein product of each one, believing that in that collection of 10,000 individual genes, there must be a signal to stimulate the immune system. The approach in these new experiments is not to test proteins one at a time but to test them 10,000 at a time. That is, we take 10,000 little test tubes containing human immune cells, put 10,000 proteins on them, and then measure the response.

Furthermore, we do not measure one or two responses with our eyes. Rather we use the power of new instrumentation to measure hundreds of responses. We obtain a highly detailed portrait of what each human protein does to each cell. We then use the power of modern computation to collect more than 2 million pieces of biological data per experiment and then scan 10 cell types, for a total of 20 million pieces of biological data. An interface allows us to quickly sort through and identify the responses we want. That is a modern experiment, and it is one that we do every day.

It took only a few months to find the immune stimulator we were looking for. Once we found that protein and described it, it was just over a year before we began treating patients. In that time we did all the steps it takes to show that the protein works safely in animals, so that we could convince the Food and Drug Administration to allow us to initiate tests in humans.

What is critical in this approach is that the problem is solved by using a system, not by relying on an idiosyncratic genius. We used the combined power of new technologies to address old medical problems. The new medicine that will come into existence will not just be more of the same chemical medicine or the better use of plant substances. It will be drawn from our own bodies.

We now have a new-found capability to alter our bodies for the better, using those self-same substances we use naturally to create, maintain, and repair our bodies. We can do better than the body was originally designed to do. This ability will provide a longer, healthier life for many people. It will allow us to repair injuries that could normally not be repaired. It will provide corrective brakes on systems that have gone out of control, be they cancer or portions of the immune system that give rise to autoimmune diseases.

The immune stimulator we discovered, when produced in overabundance, appears to cause autoimmune diseases such as lupus and rheumatoid arthritis. We now have a potential remedy. We can reduce levels of that substance by introducing into the body antibodies that specifically recognize it. It took us only a few months to find an antibody that would accomplish this task. We have now started trials for treatment of autoimmune disease using this antibody.

Thus, we can enhance or diminish normal function. These are functions that go wrong in almost everyone because as we age, all parts of our bodies wear down, regardless of our genetic differences.

By Studying One We Study All

The advances being made in our understanding of human biology extend far beyond the perimeters of our own bodies. These advances, although pioneered by our desire to treat and cure ills, are powerfully applicable to our understanding of the broad scope of living organisms. We can now begin to use the similarity of all living structures as a practical tool. We can begin to study processes that occur in bacteria, insects, or other animals and through that study gain an understanding of human function.

We can also gain insights into the functioning of other organisms. A study that is undertaken in diverse life forms is as much a study of similarities as it is a study of differences. If we apply this work not just to humans but also to agriculture, we may be able to enhance agricultural properties by using the genes in each organism in a better or more efficient way. We would not be adding genes to those that were originally there. This path will take us—as it will with humans—to truly natural science and vastly improved agriculture and medicine.

Arnold J. Levine

The Origins of Cancer and the Human Genome

 At this time, it is clear we are in the midst of a remarkable revolution in the life sciences, with the first draft of the map of the human genome representing another milestone in this journey. It is important to appreciate that this advance comes about through an extraordinary set of observations made by thousands of scientists over the past 100 or so years.

I would like to use my own field, cancer biology, to present a short history of how we have progressed in understanding cancer. In 1961, when I was a first-year graduate student at the University of Pennsylvania, I cared passionately about two things—viruses and cancer—and I particularly wanted to understand the origins of cancer in human beings, which in 1961 was something we had no understanding of at all.

Understanding the Causes of Cancer

The story began in 1911 in New York City, on the campus of Rockefeller University, where a new hospital had just been built. A young man, Peyton Rous, had just finished medical school at Johns Hopkins University and was offered a job at Rockefeller University to work for its first president, Simon Flexner. As Rous was leaving Johns Hopkins, his dean and mentor advised him to work on anything there but cancer because they had no idea what cancer was. Upon arriving in New York and settling in, however, the first thing Rous's employer, Dr. Flexner, told him was that they did not have anyone there working on cancer, so that was the subject he should work on. Those were the days when presidents of universities had real power and faculty members could not say no, which meant that, even though he did not know how to proceed, Rous began his work by focusing on cancer.

Within six months of his arrival, a chicken farmer from New Jersey walked into his laboratory with a live chicken that had a tumor on its breast. It turned out to be a tumor of the muscle cells, called a sarcoma. The chicken farmer said to Rous that his flock was coming down with tumors, and he needed to know what was going on. With that, Rous formed his first hypothesis about cancer, speculating that viruses could cause cancer in chickens. This was a convenient hypothesis because earlier work had taught us how to isolate viruses.

In 1911 viruses were isolated by sacrificing the chicken, removing the tumor, crushing the cells of the tumor, and then filtering the crushed cell extract to exclude all cells and bacteria. The fine pore size of the filter allowed only the smallest of living things—viruses— to go through. Then a clear filtrate with no cells or bacteria was used to inoculate another chicken, which would then develop a tumor. By doing this, Rous unequivocally demonstrated that a living organism was responsible for tumor development and that the organism was a virus that replicated within the tumor.

Rous's first paper, which stated that viruses can cause cancer, was received with much skepticism. Some could accept that viruses might cause cancer in chickens—that perhaps cancer is an infectious disease in that species—but the view that cancer was caused by an infectious disease in human beings was considered extreme. Today, we know of five viruses that are the cause of cancer in human beings. In fact, the most debilitating viruses in terms of numbers of humans are hepatitis B and C, which predispose infected individuals to liver cancer, or hepatocellular carcinoma. Rous's observation laid the first cornerstone in our understanding of cancer and its causes.

Causes of Cancer: Chemicals, Aging, and Genes

Now we can move forward from 1919 to the 1930s, when a team of scientists at the University of Wisconsin and another team at the University of Tokyo made the observation that coal tar causes skin cancer. Using scrapings from a chimney, they extracted the chemicals and painted them on the backs of mice, which resulted in skin cancer. This was the first time evidence had been collected that chemicals cause cancer. So by 1940 two causes of cancer had been documented: viruses and chemicals.

Then in the 1950s and 1960s a group of epidemiologists elucidated the fact that most cancers, some 90 to 93 percent, occur in elderly people. In fact, during the first five decades of life, the rate of cancer is very low. It then starts increasing at about age 55 at an exponential rate. Now three causes of cancer had been identified: viruses, chemicals, and aging.

Finally, throughout the 1950s until the 1980s, scientists started patching together epidemiological data that had been collected over the first half of the century. These data demonstrated that some cancers occur with high frequencies in certain families. Scientists speculated that even though the inheritance patterns were complex, it was clear that genes were involved.

How Genes Cause Cancer

In 1961, as a first-year graduate student, I was presented with these facts: viruses, chemicals, aging, and genes can cause cancer. But none of this told me what actually causes cancer. How is it that cancer arises from those four variables? Is it true that all four are important in humans? And could these four variables be put together in some way?

We now move into the 1970s and the revolution in molecular biology that was building on the 1953 Watson and Crick discovery of the structure of DNA. In the 1970s we learned how to clone a gene—that is, to take it from a chromosome, isolate it, and put it into a place where it could be analyzed separately from all other genes. This ability allowed scientists to test for the first time whether genes cause cancer.

Harold Varmus and Michael Bishop, who subsequently won a Nobel Prize for their work, first carried out the test. They started by studying the Rous sarcoma virus, named after Peyton Rous. What Varmus and Bishop studied were two classes of closely related viruses, some of which cause cancer and some of which do not. The class of viruses that were not cancer causing contained three genes, all active in replicating the virus. But the viruses that cause cancer in chickens, Rous sarcoma, had four genes. The fourth gene was not needed to replicate the virus, but it was needed to cause cancer. They named this gene an "oncogene," or cancer-causing gene. Thus, an amazing link between viruses and genes suddenly appeared—viruses cause cancer, genes cause cancer, and viruses carry genes that can cause cancer.

The second observation Varmus and Bishop made was equally startling and remarkable. A homologue (a similar gene) of the oncogene found in the virus is almost identical to a gene found in normal chickens. They called this gene "src" for causing sarcoma. So they postulated that there must be some difference between the normal gene and the gene in the virus, and the difference turned out to be about three or four changes, or mutations. These mutations, or mistakes, were part of the viral oncogene but were absent in the normal chicken gene. The virus had picked up or stolen the normal chicken gene, and then this gene had acquired mutations in the virus that made it an oncogene.

Moreover, carcinogens or mutagens can cause these mutations, a realization that instantaneously and conceptually tied together three of the four major cancer-causing variables identified in the past. We had chemicals that could cause mutations in genes, and the genes could be picked up by viruses. Further, the src gene found in the chicken had a homologue in humans.

Thus, not only do chickens have a normal src gene, but so do humans, meaning that mutations in that gene might give rise to human cancer. This posed a clear hypothesis for many to explore, including teams of scientists at the Massachusetts Institute of Technology, Cold Spring Harbor Laboratory, Columbia University, and the National Cancer Institute. These groups started searching for oncogenes that could cause cancer, looking for them in human tumors, and cloning those genes.

The first oncogene isolated from human cancer was named "Ras." It is a very simple molecule, like a small transistor. If you think of a pathway with on/off switches, Ras is simply an on/off switch. Like a transistor, in one form it is on and in one form it is off. The on switch means cell division is occurring. The cells start dividing and do so in an uncontrollable way. To stop cell division you turn the Ras gene off. However, a mutation in the gene causes it to turn on so that it cannot be turned off, and the cells divide in an uncontrolled fashion. Later, other genes would be found that when mutated kept the cells stuck in the on position. In many of these cases it was relatively straightforward to prove that these genes were contributing to cancer, because they could be isolated from the chromosomes of cancer cells and transferred into a mouse, and the mouse would develop cancer. Today we know of 80 to 120 oncogenes in the human genome that may contribute to cancer because of a mutation that causes cells to divide uncontrollably.

Tumor Suppressor Genes

In 1979 a second set of genes, called tumor suppressor genes, was elucidated in my laboratory at Princeton University and elsewhere. We named the first tumor suppressor gene p53. They are called tumor suppressor genes because they prevent cancer. We began to appreciate the functions of tumor suppressor genes in the 1980s and 1990s.

Let me explain the function of p53. Later this afternoon most of you will walk outside, and if the sun is still out, its rays will directly hit your skin. The ultraviolet light from the sun will react with the DNA in the cells of your skin and in some cases will cause a mutation. If those mutations happen to be in oncogenes or tumor suppressor genes and they accumulate over time, it might cause a skin cancer to develop. When those mutations occur, an alarm goes off in the cell—a signal that something has gone wrong. The p53 gene recognizes that signal, responding to a potentially dangerous cell or precancerous condition by killing the cell in a process called programmed cell death.

Thus, p53 is a sensor; it integrates signals from many places, sensing mutations in the genome. For example, a severe sunburn with peeling skin shows p53 in action causing programmed cell death of the skin. Fortunately, skin regenerates constantly, which is the reason why the strategy of eliminating some cells is a good one. We are multicellular organisms that regenerate some of our cells continuously.

But what happens if there is a mutation in p53 or it is missing? In that case, other mutations can appear in many places without being detected and corrected through cell elimination. Cancer cells actually arise at a high frequency. In the United States we know of about 250 families that inherit p53 mutations in one of their two copies of chromosomes. Anyone who inherits a p53 mutation will, with a probability of virtually 100 percent, develop some type of cancer over his or her lifetime. Today we know of 20 to 25 tumor suppressor genes that when mutated predispose us to cancer, and these are primarily the causes of inherited predispositions to cancer.

The Effects of Aging

Why is it that cancer is largely a disease of the elderly? We accumulate mutations over our lifetimes, and in order for a cancer to arise, we need five or more mutations to occur in the genetic information in a single cell of our body. For example, one of the cells of our body might, by the age of 15, accumulate one mutation; by the age of 25, a second mutation in a critical tumor suppressor gene or oncogene; by the age of 40, a third mutation; by the age of 50, a fourth mutation; and by the age of 65, a fifth mutation. Five cumulative mutations in the exact same cell of the body can activate oncogenes and inactivate tumor suppressor genes, giving rise to cancer.

Cancer, therefore, arises through somatic mutations. But what gives rise to these somatic mutations? It depends on the cancer. It could be the food we eat, the chemicals we are exposed to, or the sunlight we enjoy. Even things we cannot prevent, like gamma radiation coming from the universe, break DNA and cause mutations.

These four variables—viruses, chemicals and exposures, genes, and aging—explain the nature of the origin of cancer in human begins. Roughly 100 oncogenes and at least 25 tumor suppressor genes combined with these variables can give rise to cancer. The many combinations of mutations in these genes explain why cancer is such a diverse disease. Ten people with the same kind of cancer may have very different outcomes and prognoses. They may respond differently to chemotherapy, or one cancer might metastasize to other tissues, while another may not. Thus, cancer is a disease of combinatorics—the combinations of genes and exposures. Certainly one of the challenges we face is deciphering these combinations and finding out what they tell us and how we can respond with more effective therapies.

The knowledge of what causes cancer is important because it serves to empower us and provides the targets for cures. We can begin to understand how chemotherapy and radiation work, even though one day we must leave those therapies behind in favor of the rational drug therapies of the future that will attack oncogenes or reverse the inactivation of tumor suppressor genes.

Rational Drug Design

A whole new generation of drugs is in development that come from studying the rational basis of cancer therapy. One drug, STI-571, or

Gleevec, produced by Novartis, is now an approved drug for cancer therapy. There is a story behind the development of this drug, just as there is a story behind all the great discoveries made in cancer research over the past 40 years.

In 1961, when I was at the University of Pennsylvania, a young assistant professor, Peter Nowell, a pathologist, was studying cells from patients with leukemia, a white blood cell disorder in which cells divide in an uncontrolled fashion. He was studying chronic myelogenous leukemia, which is usually experienced for four or five years in a chronic or slowly smoldering form before a "blast crisis" occurs, an acute phase during which the cells start dividing very rapidly, leading to a rapid terminal phase. The only treatment before the blast crisis was bone marrow transplant, which has only a 40 to 50 percent survival rate.

What Nowell noticed was that all the patients he saw with chronic myelogenous leukemia had a chromosome abnormality. Normally, we have 23 pairs of chromosomes, or 46 total, but in these patients two of those chromosomes had broken and fused. This is called a translocation, and it is a mutation. Nowell hypothesized that, if every patient with myelogenous leukemia had this translocation, this would be a good correlation suggesting a causation. After Nowell published his findings, the chromosome was named the Philadelphia chromosome, because the University of Pennsylvania campus where Nowell worked is located there.

In the 1970s, when oncogenes were being discovered, in David Baltimore's laboratory at the Massachusetts Institute of Technology, a young student named Owen Witte (now at the University of California, Los Angeles) isolated an oncogene from a virus that caused cancer in rats. Because the virus was named the Abelson virus, after its discoverer, Baltimore and Witte named their gene the Abelson oncogene, or Abl.

It was later found that at the merger of the two fused chromosomes of the Philadelphia chromosome there is a mutation in the Abl oncogene. This suggested that the translocation created the mutation that caused the cancer. In fact, causality could be shown in a very clear fashion by taking the Abl oncogene from that translocation, called Bcr-Abl, and putting it in the mouse, after which the mouse developed chronic myelogenous leukemia. A number of others later uncovered the function of the gene, which is to chemically modify proteins, specifically a protein kinase, by adding a phosphate group to a protein.

In 1991 an arrangement was made between the pharmaceutical company Novartis and the Cancer Center at Harvard Medical School to work collaboratively to target oncogenes for future therapies. The goal was to find the best targets for a drug to inhibit an oncogene product, thereby curing the cancer. They decided to go after the Abl oncogene, the cause of chronic myelogenous leukemia. The function of the enzyme it produced was known, and they could purify the enzyme. They could look for drugs that would inhibit the enzyme activity and perhaps cure the disease. By 1998, they had found several good inhibitors.

In fact, the inhibitors were so good that mice with the activated Bcr-Abl translocation oncogene that had developed chronic myelogenous leukemia were given STI-571 and were cured with no side effects. Next, Brian Drucker at the Oregon Medical Center began testing STI-571 in human beings. The drug was remarkably safe with minimal side effects, and when given to people with chronic myelogenous leukemia, nearly every patient underwent a dramatic remission. Only a small number would develop resistance to the drug. These observations led to clinical trials at multiple sites and rapid approval of the drug by the Food and Drug Administration. Today, STI-571 is called Gleevec, and expanded trials are under way for other cancers, such as lung cancer.

Although this is a wonderful story about the history of oncology, we cannot yet say that chronic myelogenous leukemia can always be cured. Some patients have become resistant to the drug, which means we will need to develop a second round of drugs that are effective against other oncogenes, and we will need to use combination therapies. However, the pathway is now established, and we understand the set of genes that cause cancer in human beings. We can, in fact, strike back with rational approaches to therapy that eliminate the horrors of current treatments. And the sequencing of the human genome for the first time gives us complete information about the number of oncogenes and tumor suppressor genes we might have to deal with.

Three years ago I attended a meeting at the National Cancer Institute and asked the question: "Have we found all of the oncogenes and tumor suppressor genes in human beings, and is it time to stop looking and to start focusing on getting good drugs that make a difference?" Of the 15 scientists in the room, some said we probably had found about 10 percent of the genes, while others said we probably had found almost all of them. The real answer is that we do not know. But the beginning of the answer is in the sequence of the human genome, which will tell us how to cure the cancer that begins in our own genes.

Mary Jeanne Kreek

Gene Diversity in the Endorphin System
SNPs, Chips, and Possible Implications

 I would like to focus on gene diversity in the endorphin system, specifically natural SNPs (single nucleotide polymorphisms), custom-made chips, and possible implications. The implications are diverse, ranging from normal human physiology to diseases of the brain, with major behavioral manifestations.

The human genome is remarkably identical across humankind. Of the approximately 3 billion bases on the 30,000 to 100,000 genes, 99.9 percent are identical. That leaves 3 million, or 0.1 percent of bases, that may have variations. The most common kind of variation or polymorphism is the SNP. A SNP is one nucleotide or base that is different from the usual, or prototype, or the first that was identified and recorded. Thus, whoever (i.e., whatever human being) was sequenced first for any specific gene became the prototype, when in fact that person might have had a SNP somewhere in the gene under study.

A gene's coding region DNA codes first for a messenger RNA (mRNA), which in turn yields a protein. The rest of the gene is also important when one considers diversity and polymorphisms. Also, because variations are spread across all of humankind, most allelic variations are low in frequency, less than 1 percent. Others appear at an intermediate frequency of 1 to 5 percent. Frequencies of 5 percent or more are considered high.

Eric Lander, of the Massachusetts Institute of Technology, approximates that 1 in 346 base pairs in the coding region of a gene will have a SNP. In looking at 106 genes, he found that the number of SNPs per gene ranges from 0 to 13. He states that "in an individual human, two copies of an average gene chosen at random will differ by roughly 1 base in 2 Kilobases (Kb), corresponding to somewhat less than 1 heterozygous base in the coding region of that typical gene."

Polymorphisms, or SNPs, can be unfortunate if they result in the loss of a peptide that leads to the development of disease. But for the most part, SNPs are neither good nor bad; they are just different. Some have functional significance and some do not, and the functional significance could include different peptides or proteins coming from an altered or polymorphic gene if the change is in the coding region, or differences can lead to different levels of gene expression. These changes can lead to different responses to medications and therapeutic agents, the study of which is called *pharmacogenetics* or *pharmacogenomics*.

Variations in the Endorphin System

Two years ago we found that a genetic variant could lead to different effects of an endogenous compound, in this case a peptide that can act as a hormone or a peptidergic neurotransmitter. When such genetic variants are identified that could lead to alterations in physiology, we term this *physiogenetics* or *physiogenomics*.

In my laboratory we are particularly interested in the endorphin system. Endorphins are the natural morphine-like peptides in humans and animals. Eric Simon of New York University coined the word "endorphin" from "endogenous morphine." We now know that there are three classes of endogenous opioids or endorphins: (1) proopiomelanocortin (POMC), which yields beta endorphin, the longest of the endorphins, and also a critical stress response hormone, adrenocorticotropic hormone (ACTH), along with other interesting peptides such as the melanocyte-stimulating hormone (MSH) family; (2) the enkephalins; and (3) the dynorphins. In each case one gene yields one big peptide, which is then processed to yield many biologically active peptides in the brain as well as the periphery.

However, it was not until late 1992 that the opioid receptor was successfully cloned. We now know, as we thought was correct based on earlier selective chemical studies, that there are three different kinds of opioid receptors—mu, delta, and kappa. And although there is not a one-to-one match-up, certainly beta endorphin binds more at mu, some of the enkephalins bind at mu and delta, and dynorphins bind preferentially at the kappa opioid receptor.

Endorphins, Genes, Environment, and Addiction

We know through studies in my laboratory and many others that the endogenous opioids play a role in diverse and extremely important functions, including many involved in survival. The endogenous immediate response to pain or painful stimulus, several components of immune function, gastrointestinal function, and cardiovascular and pulmonary function are all modulated in part by the endogenous opioid or "endorphin" system. Also, we have growing evidence that mood, affect, cognition, and possibly learning and memory are modified by components of this system.

The major drug of abuse, heroin, also acts at this system. If there are alterations in levels of mRNA—that is, gene expression that may lead to altered levels of receptors, peptides, or hormones downstream—we may observe atypical or altered function of each one of

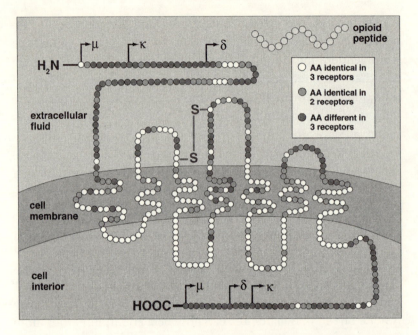

FIGURE 1 This is a diagram of the human opioid receptors, including mu, delta, and kappa. Three regions are in the extracellular fluid outside our cells. The endorphins can bind to this region. Compounds like morphine can bind to this region, to seven "transmembrane" regions, and inside the cells where the endorphin peptide-activated signal is amplified by a variety of increasingly well-understood mechanisms. Adapted from LaForge, Yuferov, and Kreek, *European Journal of Pharmacology*, 2000, with permission from Elsevier Science.

these systems (see Figure 1). Sometimes these altered functions will actually lead to or be part of the mechanisms of disease.

We have hypothesized that the endorphins may be involved in each of three major addictive diseases: alcoholism, cocaine addiction, and heroin addiction. Many laboratories continue to ask critical questions about the extent of this role and its precise mechanism. In 1964, when we (Dole, Nyswander, and Kreek) initiated our research, addictions were thought to be deviancy, personality disorders, or simply criminal behavior. We then hypothesized what is now accepted by most scientists and clinicians, that addictions are diseases of the brain with behavioral manifestations expressed in a social context.

Addictions also can be correctly defined as compulsive drug-seeking behaviors and self-administration without regard to negative consequences to self and others. The U.S. federal regulations governing entry into opioid agonist pharmacotherapy define heroin addiction even beyond this, to give a more stringent diagnosis, as multiple daily self-administrations of illicit opiates for one year or more with the development of tolerance and physical dependence as well as a compulsive drug use.

We know that in the United States over 177 million people have used alcohol, and approximately 15 million people are alcoholics. Over 26 million people have used cocaine, and 1 million to 2 million are addicted. In addition, 2.5 million to 3 million people have used heroin illicitly, and 0.5 million to 1 million have become addicted. It is intriguing to contemplate that in our nation and throughout the world where opiates have been introduced, about 1 in 3 to 1 in 5 individuals who have ever self-administered illicit heroin have become addicted to it. In contrast, only 1 in 10 to 1 in 20 who self-expose themselves to alcohol or cocaine become addicted to those substances.

We hypothesized many years ago that three separate domains of factors contribute to the development of addiction (Figure 2). We hypothesized that multiple alleles of multiple genes acting in combination would enhance either the vulnerability to become addicted when self-exposed or would decrease vulnerability or protect when self-exposed. We also hypothesized that environmental factors, from early environment, prenatal, and early postnatal, or even right up to adolescence or adulthood coupled with "set and setting" of use and other behavioral events—and also such factors as other diseases or stress—could interact with genetics to enhance vulnerability for the development of an addiction.

We also hypothesized that drugs of abuse would themselves cause changes in the brain and that these changes might result in very fundamental changes in physiology that would participate in the development and persistence of an addiction. In fact, there is experimental evidence in humans, but to a greater extent in animal models, showing that indeed all three of these factors play a role. It

Use of the drug of abuse essential (100%)

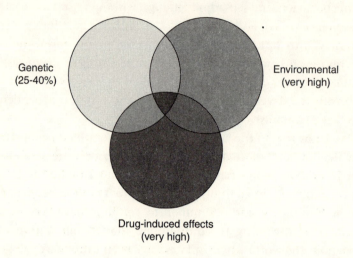

Genetic
(25-40%)

Environmental
(very high)

Drug-induced effects
(very high)

FIGURE 2 Factors contributing to vulnerability to develop a specific addiction.

is critical, however, that there is self-exposure to the drug of abuse before these events can occur and before any genetic vulnerability can be unmasked.

Addictions are complex disorders, as are many other disorders. Any genetic or inheritable contribution to the vulnerability to develop a specific addiction would involve multiple alleles of multiple genes, that is, polymorphisms (including SNPs) of multiple different genes.

Epidemiological studies have shown that there is a contribution of inherited or genetic factors to the development of alcoholism and to many other addictions, with about 25 to 50 percent of the relative risk contributed by genetic factors and the rest by environmental factors, including drug-induced changes. Recent epidemiological studies by Ming Tsuang of Harvard in over 3,000 monozygotic and dizygotic male twins have shown that of all the addictions heroin addiction has the largest amount of unique genetic variance, that is, unique to opiate addiction and abuse, while also having the lowest amount of shared genetic variance contributing to the addiction.

Of the three drugs of abuse, heroin is a known depressant. It acts primarily on the endogenous opioid system and very specifically on the mu opioid receptor system. It also affects the dopaminergic system and other neuropeptides and neurotransmitters. Cocaine is a stimulant, and it acts primarily in the dopaminergic system but also the serotonergic and noradrenergic systems, the three major intermediate speed neurotransmitter message systems in the brain. It does so by blocking the normal presynaptic reuptake, causing an increase in function or activity level of these transmitters. We have ample proof that cocaine also profoundly affects the endogenous opioid system. Alcohol is both a stimulant and a depressant. Its primary actions alter both the dopaminergic and the opioid systems as well as other neurotransmitter systems.

In our earliest studies we defined the nature of heroin addiction and found that the heroin addict typically self-administers this short-acting opiate three to six times a day, initially to get euphoric or "high" but later, with the development of tolerance and physical dependence, to prevent the onset of opiate withdrawal symptoms (see Figure 3). We know from animal modeling that administration of a short-acting opiate intermittently in this way causes profound disruptions in gene expression and other neurochemical events.

In contrast, methadone is a mu opioid receptor selective synthetic opioid that is long acting in humans, with a half-life of more than 24 hours, as opposed to heroin, and its major morphine metabolite, which last only 3 minutes and 4 to 6 hours, respectively. We found in 1964 that heroin addicts treated with this medication became normalized and experienced no "high" or withdrawal symptoms but remained in a normal behavioral and functional state. Today, over 180,000 former heroin addicts in the United States are in successful methadone maintenance treatment, which combines pharmacotherapy with counseling. We later learned in animal studies that gene expression is not altered by steady state infusion of methadone, whereas there are profound alterations during intermittent morphine administration.

In other studies of drugs of abuse, we have learned that the "on/off" effects of these drugs profoundly alter levels of expression of

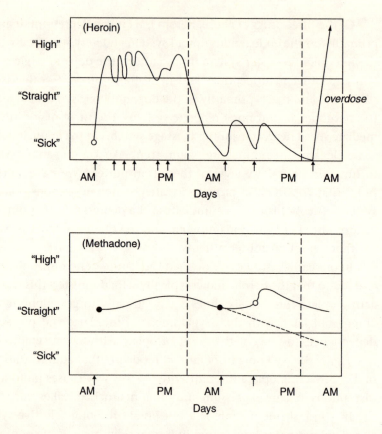

FIGURE 3 Impact of short-acting heroin versus long-acting methadone administered on a chronic basis in humans (1964 study).

many specific genes, receptor-mediated events, as well as integrated
systems physiology and behaviors. We also hypothesized many years
ago that environmental factors such as stressors could alter
responsivity to drugs of abuse and, moreover, that stressors might
contribute to the persistence of and relapse to drugs of abuse. Such
atypical or altered stress responsivity in some individuals might exist
prior to the self-administration of addictive drugs and actually lead
to the acquisition of drug addiction (see Figure 4). Genetics and

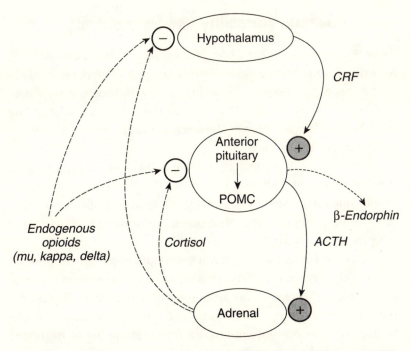

FIGURE 4 Interrelated roles of the hypothalamic-pituitary-adrenal axis and the endogenous opioid system in the biology of addictive diseases. We know in humans with respect to our stress-responsive, hypothalamic-pituitary-adrenal axis that our major stress-controlling hormone, CRF, is produced from the hypothalamus. It acts at our anterior pituitary to cause production and release of beta endorphin, as well as a major stress-responsive hormone, ACTH. This beta endorphin has the longest half-life, circulates in blood throughout the body, and, in fact, is also made in the gastrointestinal tract and lymphatic system, as well as the brain. ACTH acts on the adrenal cortex, causing the production of cortisol (our major stress-responsive steroid), which acts in a negative feedback mode to decrease the production at both the hypothalamus and the anterior pituitary of our stress-responsive hormones. Drugs of abuse profoundly alter this system by altering gene expression and then related neurochemical events.

environment might each contribute to this atypical stress responsivity. In subsequent studies it was shown that, in fact, stress induced in an animal can cause alterations in gene expression, neurochemical events, and behaviors, including self-administration of drugs of abuse.

Genetic Variability and Susceptibility

All of these changes occur when gene expression levels are changed. Knowledge about the genome allows us to appropriately use modulation of gene expression in the management of disease as well as to understand many diseases that may be the result of inappropriate modulation of these systems. However, actual genetic variations also are important to consider.

Some of the individual genetic variability and susceptibility to develop persistent addiction may in fact be due to polymorphisms of multiple genes. Since the primary site of action of opiates is the mu receptor, we postulated that polymorphisms of the mu opioid receptor might contribute to this role.

Polymorphisms that might not contribute to an addiction sometimes could serve as markers with which to scan the human genome. Individual differences and responses to endogenous opioids—physiogenetics—or opioid medications—pharmacogenetics—could be important in our response to our own endorphins or treatment agents. We decided to focus on the mu receptor and the coding region of this receptor because of its possible role in variations in susceptibility to addiction. What we found initially were five different SNPs in the human mu opioid receptor gene, of which three resulted in amino acid changes. With amino acid changes, one could postulate a potential change in function. It was of particular interest to us that two of these SNPs had a very high allelic frequency of 10.5 percent for the A118G allele and 6.6 percent for the C17T allele. In our first study of 152 well-characterized subjects, which was a small number, we found several persons with homozygosity as well as heterozygosity for each of these two common SNPs.

Further studies of the SNPs and other polymorphisms of the mu opioid receptor have yielded 10 polymorphisms in the coding region alone and also 6 polymorphisms in the gene of the kappa opioid receptor, including a high-allelic-frequency repeat-type polymorphism that may be important for gene expression. Combined with work done by others, we now know there is considerable diversity in this "endorphin" or endogenous opioid system, with identified

2 delta opioid receptor SNPs, 6 kappa opioid receptor SNPs, and 15 mu opioid receptor SNPs in the coding region alone and many more outside it. These are higher frequencies of SNPs in the coding region of the mu opioid receptor than have been found in the most common or abundant polymorphisms found by Lander in his study of 106 genes.

Is there any association between either of these two polymorphisms and opioid dependency? The answer is *no*. The C17T allele comes close to being associated with opioid dependency ($p = .05$) but not close enough. Further studies may yet reveal such an association. In addition, across all ethnic groups combined, the A118G allele had no association either with opioid dependency or lack thereof.

So in collaborations between our laboratory and that of Lei Yu (University of Cincinnati), we have gone on to ask this question: What may be the impact on both binding of various endorphins, as well as exogenous opioids, and on the amplification signals coming from that binding? We then went on to ask about one of the most important signal transduction mechanisms of the mu receptor, which is a G protein activated inwardly to rectify the potassium channel, or GIRK.

Since these two mu opioid receptor SNPs are of very high allelic frequency and may be very important for addiction, endogenous responses to pain, and stress responsivity, as well as multiple physiological functions, we wanted to be able to more rapidly identify these SNPs. We (LaForge, Mirzabekov, and Kreek) therefore created a custom microarray (or SNP Chip), that allows us to put innumerable polymorphisms on one glass slide array for comparative purposes (see Figure 5). Using this gelpad microarray technique, we can identify these or any other important polymorphisms that could be important to understand how each of our own bodies responds to our own endorphins and possibly in the future to some medications that could be effective in helping any one of a number of disorders, if the medication were targeted to the system.

FIGURE 5 Application of oligonucleotides to activated gel pad microarray.

Cautions

It is clear that gene diversity may be involved in the vulnerability to develop addictions or other disorders. However, in pursuing this possibility we must consider the privacy and confidentiality of those who are identified as "vulnerable" to prevent any kind of undesirable outcomes or reprisals.

I cringe when I think about what could be the impact if we are able to identify and then develop a panel that defines the alleles that may enhance the vulnerability to develop alcoholism, heroin addiction, or other chemical dependencies. In addition to the economic concerns, we have to be deeply concerned about the stigma that exists for people with these disorders. We have the responsibility to consider what we will do with our own information and consider the concepts that have been developed in other domains when confidentiality is essential.

Who should have access to this information? Somebody who has the "need to know" because they are going to help us medically with something we desire to be done. Who has a right to know this (i.e., our own genetic) information? It should be somebody to whom we have given our informed consent. It is our decision.

Barbara A. Schaal

Genomics and Biotechnology in Agriculture

 Agriculture, like medicine, is rapidly changing because of advances being made in molecular biology, particularly in the fields of genomics and biotechnology. However, although the application of genomics and biotechnology to agriculture has much potential benefit for the human population, these technological advances have raised widespread debate about a number of scientific, ethical, and social issues. In fact, the current public debate about the application of biotechnology to agriculture is extremely active and visible because the agricultural varieties produced by direct genetic modification are now widely planted and products from these genetically modified organisms (GMOs) are widespread in the marketplace. The debate is also international in scope, and it is often exceedingly bitter, with episodes of test plants being uprooted in fields and arson occurring in laboratories.

But just as genomics provides the basis for future advances in medicine, the application of genomics holds great promise for agriculture. Genomics will provide improved varieties of crops for the U.S. market as well as entirely new products for our economy, with potentially reduced environmental consequences, such as reductions in agrochemical use, including pesticides, herbicides, and fertilizers. One of the most important uses of biotechnology is the application of genomics to agricultural issues in the developing world, particularly in tropical regions where most of the world's poor reside and where continual challenges are presented by food shortages. Thus, biotechnology can contribute to the food security and nutrition of the world's poorest people, and, in fact, because good health is predicated on adequate nutrition, if the poor are to benefit from modern medicine and if medicine is to be ultimately successful in the developing world, the human population must be well fed and nourished.

I would like to outline some of the work currently in progress regarding plant genomics and discuss how this work contributes to international efforts in agriculture; I will then briefly compare and contrast traditional plant breeding with the production of new plant varieties by genetic engineering; and finally, I will outline the controversy surrounding GMOs, using examples related to cassava, an important tropical subsistence crop.

What is biotechnology? The term itself can cause confusion. Some define the new biotechnology as "the use of biological materials, cells and molecules, to solve problems or to make useful products."[1] Many aspects of biotechnology are included in this broad definition, such as genomics, genetic engineering, and plant tissue culture. It is also important to point out that not all aspects of biotechnology are controversial—the use of genomic markers to produce a new variety of tomato by traditional breeding or the use of tissue culture to grow orchids does not create concern. In addition, the use of tissue culture to replicate, or clone, an apple variety is generally not considered controversial. However, when we cross kingdom lines to clone sheep and pigs, public concern becomes evident. Even so, the greatest area of concern at this time remains direct genetic modification—the insertion of a gene from one species

Box 1
Concerns about genetically modified crops

- Food safety—allergens
- Escape of genetically modified organisms into the environment
- Contamination of nongenetically modified crops or native species
- Production of "superweeds"
- Effects on nontarget organisms (e.g., monarch butterfly)

into the genome of another—that is, genetic engineering, or the production of GMOs (see Box 1).

Plant Genomics

Parallel to the Human Genome Project, projects have been completed to directly sequence the entire genome of several plant species, including rice—the most important crop worldwide—and the model plant *Arabidopsis thaliana*, a member of the mustard family. Other plant species have extensive physical maps of their genome under construction using a variety of polymorphic markers, such as microsatellites, the highly variable markers used in DNA fingerprinting. Genomic mapping studies can be used to identify genes of agricultural importance, just as we have seen for cancer-related genes in the human genome. Plant scientists are specifically interested in the number and location of genes that confer resistance to pathogens—that is, genes that are involved in disease resistance—and in genes that convey tolerance to drought, temperature, or other environmental stresses that cause an estimated $500 million of lost crop production per year. This would include genes that confer tolerance for heavy metals or other pollutants and, of course, genes that increase both the yield and nutritional composition of plants. One area in which agricultural genomics directly differs from human genomics is its application to direct breeding. Although the thought of breeding humans is repugnant, we breed plants and animals all the time. Genomic information—the association of mapped markers with a desirable trait—can be used to increase the efficiency of traditional plant and animal breeding in a process known as marker-based selection.

A major issue in traditional plant breeding is identifying suitable traits for crop improvement. Where do we turn to find the genes that add value, such as disease resistance or nutrition, to crop or animal varieties? One application of genomics involves characterizing the degree of similarities and differences among collections of plant varieties that are used as the basis of breeding programs for crop improvement. Genomics can help explore whether a collection of varieties represents all the variation within a crop or whether crop improvement efforts are inevitably doomed to failure because the necessary traits are not available (see Box 2). Yet another use of genomics, only recently widely appreciated for its importance in providing new crop traits, is understanding the origin of a crop. Wild ancestors usually contain 75 percent more variation than the derived domesticated crop, and included in this natural variation may

---

Box 2
Traditional plant breeding and genomics

Traditional plant breeding
  • Crossing of plant varieties or related species to introduce new genes (traits)
  • Genes are from closely related species
  • Many genes are introduced
  • Selective breeding over generations
  • Process is slow, often requiring many years

Genomics
  • Genome mapping: identification of genes associated with crop productivity
  • Marker-assisted selection
  • Characterization of plant genetic resources
  • Origin and domestication of crops
  • Protection of plant varieties

be traits that can dramatically improve crop varieties. Finally, genomics can provide genetic markers to identify specific varieties of crops or a specific animal, while a genetic fingerprint can be used to identify a variety and thus protect the work of breeders from unlawful use. Fingerprints have also been used in plant forensics and in tracing the origin of particular varieties or breeds and, in a sad commentary, in efforts to ensure that the animals judged in 4-H competitions are the same ones that a child began raising.

Use of Genomics to Study Cassava

Examining some of the uses of genomics in the study of cassava (*Manihot esculenta*) is instructive. Cassava is also known as yuca, and in the United States it is known as tapioca. Cassava is the primary source of carbohydrates for more than 600 million people in tropical regions, mainly in Africa and South America, although its use in Asia is rapidly increasing. Cassava is grown for its starchy tubers, which are most often used to prepare farina or flour, and it is the primary source of carbohydrates in sub-Saharan Africa. It ranks sixth in overall world production. Yet despite its clear importance in feeding the developing world, cassava has been considered an orphan crop or one that is not commercially viable. Until recently it was grown primarily by the poorest of subsistence farmers, with minimal local or international trade. Because there has been little economic incentive for development, the crop has received much less attention from plant scientists than mainstream crops such as corn, soybeans, rice, or wheat. However, this picture has changed because of the efforts of such organizations as the Rockefeller Foundation, and cassava is now generally acknowledged as an important crop that holds a central role in the enhancement of food security in the tropics.

Efforts to improve cassava make use of genomic research. In fact, an international effort is under way to map the genome of cassava in order to identify genes of importance in enhancing food security and nutrition and to expedite traditional breeding. Other studies have examined the genetic basis for future crop improvement by cataloging the diversity of germ plasm or variety collections. Finally, genomic studies of the origin and domestication of cassava have yielded important new traits for breeding efforts.

As part of the international effort to improve cassava, collections of varieties at international agriculture stations were surveyed to provide new traits for breeding. It became clear that most collections of varieties were assembled for flour production and that there was little variation in other agronomic traits. In order to advance cassava development, new traits were needed. A two-pronged approach to provide more variation for the cassava breeder is being used. One part of the approach is to find additional natural variation in the plant species itself, while the second is to genetically engineer new traits into the crop.

As recently as 1990, the wild species that gave rise to cassava— the wild progenitor of cassava—was unknown. Dramatically different hypotheses were set forth about cassava's origin, one localized to Mexico, the other a single wild progenitor in Brazil. But because traditional methods of morphological analysis were unable to resolve the origin of the crop, an arsenal of genomic information was employed. One such genomic study used a combination of DNA sequences from two different genes in the cassava genome.

Genomics can be used to further refine the hunt for suitable traits and genes. Where in the range of this wild species was the plant domesticated? The wild progenitor, *M. flabellifolia*, occurs in the transition zone between the Amazon forest and the *cerrado*, a dry savanna region of Brazil along the southern border of the Amazon region. Ken Olsen, a former graduate student at Washington University, conducted this work using variations in microsatellites (DNA fingerprint loci) as well as DNA sequences of various genes, in this case an intron of a metabolic enzyme, glyceradehyde-3-phosphate dehydrogenase. The intron is a noncoding sequence that accumulates mutations rapidly and provides fine-scale resolution. Using these data, we can see that the populations of *flabellifolia* only in one part of its range contain variants found in cassava, providing strong evidence that domestication occurred in this region of the Amazon basin. In fact, cassava is part of an agricultural complex. Jack beans, chili peppers, and peanuts were all domesticated in the same region.

This information on the precise geographical location of domestication can be used to guide the search for new traits and to target geographical regions for conservation efforts and provides a good example of genomics informing conservation.

In the case of cassava, we can also look at the efforts of the traditional people of the Amazon as they selected natural variations in the crop for their own use. For many crop species, such as corn or wheat, varieties involved in the early stages of domestication are lost. This is not the case with cassava, which provides a unique opportunity to look at early varieties of the crop and to obtain information on the process of plant domestication.

Throughout the southern part of Brazil, large fields of cassava are grown for flour and starch in a manner similar to the way we grow crops in the United States. In the Amazon, however, where cassava was first domesticated and where there has been a long history of association between the crop and humans, we find a very different situation. The crop is grown in small intercropped fields with many other crops. There is a diversity of uses, with some varieties used for flour, some for boiling the roots, some for their green leaves, and some for a fermented drink. We can see the astounding diversity of the crop in the Amazon in the shape of the root, in the deposition of starch, and in the color of the root. One of the color variants, yellow, has high concentrations of beta-carotene, a significant finding because a major health problem in the tropics is lack of vitamin A, resulting from a deficiency of beta-carotene in the diet. Lack of vitamin A causes night blindness, with hundreds of thousands of children, particularly in Southeast Asia, affected. These types of variants, already in the crop, are extremely important, and because they are integrated into the plant genome, it will be easier to incorporate them into other varieties of cassava, either by traditional breeding efforts or genetic engineering.

Introducing New Traits into Crop Varieties

So how do we transfer genetic traits into crop varieties and how do crop breeders develop new varieties? Modification of plants for human use is hardly new. Humans from the earliest times have sought to use plants and animals for their own benefit. The earliest farmers in the Middle East, China, Mexico, and Africa began to grow plants they had collected for food or fiber first in the wild. They chose plants with traits that they favored, the individual with bigger seeds or with longer and tougher fibers, and they used the seeds of these plants to begin the next generation. Thus, slowly, over many generations, differences accumulated between the domesticated crop and its wild relative.

In some cases, such as corn, the process so changed the crop that the wild parent species of the crop is no longer obvious. Think about cauliflower—there is nothing that looks like it in nature. Thousands of years ago early farmers intercrossed plant species growing in their local region to produce new varieties of crops, and when the new varieties were useful, they traded seeds and animals over vast geographical scales. In fact, in the development of some crops such as wheat or kale, different species have been crossed in order to incorporate genes from one species into the genome of another (see Figure 1). Thus, the concept of using genes from different species as a basis for crop improvement is hardly new, while interestingly one of the major concerns about biotechnology has been the introduction of foreign genes into a species.

Crop breeders follow the same principles today as did those early farmers, although they use genomic information—the association of a trait with a marker—to raise the efficiency of breeding. In the example of traditional breeding shown in Figure 1, two lineages are crossed, and the progeny are examined for desirable and undesirable traits. The best-suited plants or animals are then used to start the next generation, and the process continues for what can be many generations.

What are some of the characteristics of traditional crop breeding? First, a source of new genes or traits is obtained. The source in traditional breeding comes either from other varieties of the same crop or from wild relatives or closely related species. Traditional crop breeding is an inexact science, and many genes beyond those for the selected trait, such as disease resistance, are introduced,

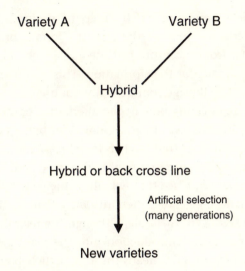

FIGURE 1 Traditional agriculture.

sometimes even whole sections of chromosomes that often may contain some genes that produce an undesirable trait (such as early dropping of seeds) or that impair crop development (genes of opposite effect that are linked). After the initial cross, the progeny and their progeny are crossed repeatedly over several generations in order to eliminate undesirable genes and to concentrate desirable traits. The process may be very slow, particularly in the case of perennial crops such as bananas or cassava for which the generation span—the time to first flowering—may be several years. Even in annual crops the process is slow. This is not, of course, to suggest that traditional breeding is unsuccessful. All of our crops are based on traditional plant breeding, including those used in the United States as well as those of the green revolution, and this has increased the yield of important crops such as rice in Asia. Regardless of future technological advances, traditional plant breeding will be an important source of new varieties or will provide the background stock for new crops produced by genetic engineering. In fact, traditionally bred varieties of crops are extremely important in this age of GMOs. The choices of which background and which variety to use for genetic transformation are critical. Some of the earliest efforts at producing GMO crops were far from successful because a relatively poor variety was chosen as the stock for transformation. This occurred in tomatoes, making the GMO lineage commercially nonuseful.

Genetic engineering presents an alternative to traditional plant breeding. Using the techniques of molecular biology, a single gene that codes for a desired trait, such as insect resistance, increased protein content, or tolerance to drought, is isolated and then combined with a promoter sequence that will allow the gene to be expressed. This combination of genes is then introduced directly into the plant genome. The concept is simple, although the techniques are technologically complex. Introduction into the plant genome can be accomplished by physical means through particle bombardment or can be done biologically. The bacterium *Agrobacterium tumefaciens*, which causes crown gall disease in plants, is used to introduce foreign DNA. Leaf disks are made of the target species—the plant species that will be altered, genetically modified, or transformed. The leaf disks are incubated with the bacteria, which infect the cells of the leaf disk—the plant cells. The bacterium contains a plasmid, a circular piece of DNA that holds the gene and promoter sequence. When the bacteria infect the plant cells of the leaf disk, in some cases the plasmid DNA with its genes is carried along and is inserted into the genetic material of the plant. These genetically transformed cells are then grown by tissue culture into whole adult plants that now contain the foreign gene and can produce seeds by standard crossing or the pollination of one plant by another. Thus, the plants can replicate, and the seed companies can build up stocks of seed that will produce new plants that will also have the new inserted gene.

How do plants produced by genetic engineering differ from those produced by traditional breeding? First, the process is highly specific. Only targeted DNA is introduced into the plant—that is, specific genes are added to the target species, as opposed to many genes introduced by traditional breeding. Second, genes can be introduced from a wide variety of organisms. Traditional breeding is limited to closely related species, within the same plant genus for the most part. Genetic engineering can use genes from across kingdoms, and plants can be engineered to contain genes from bacteria, fungi, and animals, which in turn can dramatically increase the range of traits that a plant can express. Plants are currently being engineered to serve as factories to produce useful compounds that are unlikely to occur in nature, such as pharmaceuticals, plastics, and human vaccines. A final difference between traditional breeding and genetic transformation to produce new varieties is the time involved. Breeding studies take years, while genetic transformation can be accomplished relatively quickly and more efficiently. In a perennial crop such as cassava or bananas, it takes a long time to conduct breeding studies, and because of generation time it also requires vast amounts of space and labor to grow large the numbers of individuals needed to be able to screen for selected traits. Genetic transformation occurs in the laboratory and only after it is successful are plants transferred to the greenhouse and ultimately the field.

Potential Benefits of Genetically Modified Crops

Advocates of biotechnology emphasize several advantages. By introducing insecticides that are directly produced in the plant, their repeated application can be reduced. Likewise, the nutritional content of food can be increased, novel compounds such as pharmaceuticals and vaccines can be developed, and crop yields can be stabilized by increasing resistance to drought, temperature, salinity levels, and pests. There are, of course, several famous examples of genetically modified plants. *Bt* corn is a well-known and controversial example that gets its name from *Bacillus thuringensis*, a common soil bacterium that produces an insecticide in the form of cry proteins. There are several different varieties of *Bt* corn, and they differ in the specific cry protein used and where in the plant it is expressed. The bacterial gene for the cry protein is engineered into corn to protect the plant from the European corn borer, a severe corn pest in the United States. *Bt* is considered a natural insecticide and is used as such by the organic farming industry. The controversy that surrounds *Bt* corn occurs in two main areas. One is the killing of nontarget organisms, such as monarch butterfly larvae. Some studies have shown that monarch larva die when fed *Bt* pollen, although other studies of swallowtail butterflies show little effect. The other area of concern regarding *Bt* corn is the development of resistance to *Bt* by insects. The organic farming industry is concerned about this resistance because *Bt* use is an important component of its farming practices.

A less controversial example of genetic engineering is the development of rice to express high levels of beta-carotene, the precursor to vitamin A. This is an encouraging use of biotechnology with great potential for improving the health and nutrition of the poor, particularly in Asia, where many children are fed only rice and develop symptoms of vitamin A deficiency, including blindness and retardation. Rice grains that are engineered to express beta-carotene have a clear yellow color, with the beta-carotene genes coming from narcissus plants. Intellectual property rights issues surround the development of golden rice, with more than 70 different disputes involved, as various companies, countries, and individuals make claims to biological materials or the genetic processes used in the rice's development, from the choice of plant variety to the techniques of genetic modification. Usually such claims are resolved by the payment of royalties. In the case of golden rice, many of the intellectual property rights claims are being waived as a gesture of goodwill. But these issues will be a major factor in the development of new varieties, particularly those intended for the developing world, where financial resources are slim. Finally, plants can be engineered to absorb pollutants, which can be an important component of environmental remediation. For example, tobacco plants can absorb heavy metals, mercury, copper, and lead. Under development are plants that absorb pollutants. These plants are then harvested and properly disposed of, reducing the level of pollutants in the soil.

Concerns About Genetic Engineering of Crops

Although these developments clearly have their beneficial aspects, genetic engineering has come under close scrutiny and criticism, as illustrated by the *Bt* corn example. The issues surrounding biotechnology are extremely complex, and many of the criticisms, concerns, and fears are scientific in nature. Proponents of biotechnology argue that crop varieties produced by biotechnology are carefully regulated by the U.S. Department of Agriculture, the Environmental Protection Agency, and the Food and Drug Administration and have undergone detailed testing far in excess of traditional crops. Critics of biotechnology point out that long-term effects have not been monitored and that the effect on the food supply is unknown because foods produced by GMOs are not labeled. But many concerns about biotechnology also are founded on much broader social issues, such as the ethics of inducing genes into vastly different species, the control of agriculture by large multinational companies, and our right as consumers to know what is in our food.

What are some of the specific scientific concerns? One concern involves the safety of food, in particular the possibility of the introduction into food of a foreign protein that may be allergenic to some members of the public and that is an unsuspected food component, based on consumers' experience. The example often cited is the well-intentioned effort to introduce a brazil nut protein into soybeans to enhance protein quality, even though some people are highly allergic to this protein and would not expect to encounter it in their food. This product was never developed. Another concern involves the escape of genetically modified organisms into natural environments. This is an issue particularly in marine organisms such as fish or shellfish. What would the consequences be if salmon twice as large as normal began to reproduce in a natural ecosystem? We simply do not know. Also of concern is possible contamination by GMOs—the mixing of seeds in the food supply or in seed lots sold to farmers for planting. A good case in point is a recent story about GMO corn that was approved only for animal consumption being found in taco shells.

Another concern is hybridization of GMOs that affect the biology of native species and have negative effects on nontarget organisms—for example, the killing of monarch butterfly larvae by pollen from *Bt* corn. Other issues include the development of disease resistance and the use of antibiotic markers in developing GMOs. These issues must be addressed through scientific study and the determination of relative risks. This is a complex mix of questions that involve many different species and for which there will be no single set of answers. Instead, answers will be specific not only to the issue but also to the species and the location. For example, in the case of *Bt* corn, contamination by genetically engineered corn of wild progenitors of maize is an issue in Mexico where the progenitor grows, but it not an issue in the U.S. Midwest, which has no close relatives of corn. The effect of *Bt* corn on butterfly populations depends on which butterfly species is being considered and which genetic construct of corn is planted as well as on whether or not the larvae are eating at the same time that pollen is being shed, which will vary across the country.

Conclusion

How do we deal with these issues of biotechnology as a society? Doing nothing means that we forego employing a powerful technology, one that holds great promise for improving human health and nutrition, developing sustainable agriculture with reduced environmental consequences, and developing new products and compounds that provide economic growth. On the other hand, it is clear there are a number of scientific issues that must be addressed. In order for agricultural biotechnology to reach its full potential, those directly involved in biotechnology must listen to the public debate and concerns, and careful scientific studies that are open to scrutiny and discussion must be conducted. Finally, the public must be an informed participant in the process.

Acknowledgments

The author acknowledges grant support from the National Science Foundation.

Note

1. H. Kreuzer and A. Massey, 1996, *Recombinant DNA and Biotechnology*, American Society for Microbiology, Washington, D.C.

Part III

Exploring Human Variation

Understanding Identity in the Genomic Era

Rob DeSalle

Introduction

> Knowing yourself means knowing what you can do;
> and since nobody knows what he can do until he tries,
> the only clue to what man can do is what man has
> done. The value of history, then, is that it teaches
> what man has done and thus what man is.
>
> —R. G. Collingwood, *The Idea of History*, 1946

The papers in this section are about history. History is hidden away in our genes and in the medical and family histories used to examine disease. The history of past social mistakes concerning genetics should instruct our future. Our view of the human genome is better examined in the context of history.

History is frequently hidden from plain sight. Discovering the events of the past is often a detective story. So it is with the history of human movement across this planet. Since the 1970s, when geneticist Richard Lewontin and others articulated the great paradox of human variation—that there is more variation within populations or ethnic groups than between them—we have been aware of the biological unity of all humans as a species. Current sequencing efforts and studies of single nucleotide polymorphisms, or SNPs, have verified that there is no hierarchical structure of the human population or among ethnic groups. This result has prompted the former president of Celera Genomics, Craig Venter, to assert that "the only race is the human race." Indeed, one of the first great results of the Human Genome Project is the rejection of the concept of race as a biological phenomenon.

In fact, as more and more sequence data accumulate, while the gene genealogy trees being generated are branching, there is no part of the hierarchy that can explain existing geographic or skin color "groups." Any structure that exists in the Y chromosomal, mitochondrial, and X chromosomal trees in the literature are eroding as more chromosomes are examined and more humans are added to the studies. So how can recent studies make claims of human migration patterns? The answer is good detective work. The footprints of past human migration are hidden away in specific genes in our genomes because of the way these genes are passed from parent to offspring. Like any good detective, scientists who examine these questions about human population movement must use the right tools for the job.

In the case of human migration, if you want to follow how female humans have dispersed over the planet, you need a tool that follows just female lineages. Likewise, if you want to examine male dispersal, you need a tool that follows just male lineages. Fortunately for human evolution detectives two tools exactly suited for this purpose exist—mitochondrial DNA (mtDNA) for following females and Y chromosomes for males. Unlike the rest of the genome in general, mtDNA and Y chromosomal DNA do not recombine, and are passed on in a clonal fashion. That is, tracing the history of these DNA sequences is relatively straightforward because recombination does not meddle with the inheritance of sequences in these stretches of DNA. Biologist Douglas Wallace of Emory University has used both tools in his detective work on human migration patterns and has uncovered increasingly fine detail on how human male and female lineages have moved over the globe. In thinking about the inferences from this work, we must resist thinking that mtDNA and Y chromosomal patterns reflect the current hierarchy of humans.

The patterns obtained from these markers are merely the imprints of history hidden in our genomes. These two marker systems make up no more than 51 Mb of DNA, which means that they reflect a little more than 1 percent of the entire genome of a human. This helps us understand that these markers are only a small part of our genomes.

Variation is another source of information hidden in our genomes. In this respect talking about "99.9" appears to be one of the major mantras of the Human Genome Project. This number refers to the percentage of the genome that, on average, is similar from one randomly chosen human to the next. This also means that, from one human to the next, 1 in every 1,000 bases is different; taken in the context of the entire genome, this means that there are nearly 3 million differences in the genetic code between any two randomly chosen humans. Most of these changes are silent and have no impact on the overall outer appearance, health, or behavior of humans.

By looking to our more recent past, we can also see the impact of genetic research on society. Many have argued that genetics affords a platform for improving society at large. Others have discussed at length the dangers of placing too much emphasis on genetics in society. Nonscientific application of genetics—the eugenics movement in Europe and North America, and Lysenkoism in the former Soviet Union as examples—had horrific consequences. Many academic institutions in Europe and the United States were supportive of the eugenics movement. The American Museum of Natural History, the host of the conference that produced this volume, hosted two international eugenics conferences in the 1920s and 1930s. The plain fact of the matter is that the eugenics movement was not science but rather social policy hiding behind pseudoscience. In this section, Daniel Kevles, a historian at Yale University, places the specter of eugenics in a modern genomics context.

The new eugenics might be most evident in our increasing capacity to enhance our genetic makeup. Historians David and Sheila Rothman discuss this issue in their description of genetic enhancement. Both the Rothmans and Lee Silver, a biologist, have suggested the possibility that only the wealthy will have access to enhancement technology. This seems to be the greatest fear of most social scientists, ethicists, and biologists who work in this area.

How we view our past as well as our future is critical to our interpretation of sequencing the human genome. Many of the contributors to this book were adamant that we are not headed for a social version of eugenics like that which arose in the early part of the twentieth century. But we must always be mindful of the capacity to abuse or misuse any biological information, and we should not fool ourselves into thinking that some people will not attempt to misuse these technologies. To combat such attempts, we must make the public aware of the history of our species, and knowledgeable about the future beneficial applications of information derived from the human genome.

Douglas C. Wallace

Using Maternal and Paternal Genes to Unlock Human History

The Human Genome Project has told us much about the structure and function of the human genome. However, to obtain essential insights into our origins and the causes of degenerative diseases, we must determine the nature and extent of the genetic variation of our genomes.

There are now over 6 billion human beings on the planet, distributed from the Arctic Circle to Tierra del Fuego. They exhibit striking differences in physical features, indicating adaptation to different environments. With the genomic revolution new tools have become available to study human diversity at the DNA level. With these tools we have been able to reconstruct human history with a surprising degree of clarity.

While the fields of archeology and anthropology have provided vital physical and cultural information about our species, many questions about our origins remained unanswered and unanswerable—until the Human Genome Project changed all that.

DNA is a historical molecule that retains information about the history of life on earth from its origin and evolution billions of years ago up to the recent origin and radiation of our own species, *Homo sapiens*. We are the products of all the evolutionary experiments that have occurred since the beginning of biological time. This wealth of historical information has been methodically passed down from generation to generation through DNA replication. Thus, the information in our genomes carries with it a record of our prehistory. By comparing DNA sequences of different species and individuals, we can read the history of man and woman.

DNA has the capacity not only to encode and transmit information through DNA replication but also to accumulate mistakes, known as mutations. Mutations can change any base to another and can result from errors in DNA replication as well as DNA damage. Hence, mutations occur randomly and accumulate steadily with time.

There are two broad categories of mutations: those that change a DNA base, which in turn alters a cellular function, and those that change a base without altering a genetic function. These latter mutations are assumed to be genetically "neutral" and accumulate roughly proportional to time. Therefore, the number of mutational differences between the same DNA segments of two individuals (the nucleotide sequence divergence) is roughly proportional to the time since they shared a common ancestor.

As a result, individuals who are closely related have very similar DNA sequences, whereas those who are more distantly related have very different sequences. The time since two individuals shared a common ancestor is called the coalescence time.

Biparental Autosomes and the Uniparental Y Chromosome and Mitochondrial DNA

Our genome is encompassed in 46 chromosomes plus the mito-chondrial DNA (mtDNA). The 46 chromosomes are composed of 22 pairs of autosomes and a pair of sex chromosomes: XX for females and XY for males. The paired autosomal chromosomes are the bearer of the classical human genes, one copy inherited from the mother and the other from the father. While there is a tremendous amount of historical information contained in the autosomal DNA, it has been less useful in reconstructing human origins because this infor-mation is regularly garbled by recombination. When forming the sex cells to generate the next generation, each pair of autosomes aligns and the maternal and paternal chromosomes reciprocally exchange bits and pieces of their DNA molecules. In this way the autosomal genetic information of different lineages becomes scrambled over time, and direct lineal associations become difficult to decipher.

By contrast, the Y chromosome and the mtDNA are inherited from only one parent, the Y from the father and the mtDNA from the mother. Hence, these genomic elements cannot pair with com-parable elements from the other parent and thus do not undergo recombination. As a result, they accumulate mutations sequentially along radiating paternal or maternal lineages, respectively, permit-ting reliable reconstruction of patrilineal and matrilineal relation-ships.

The Y chromosome determines maleness. A father transmits his Y chromosome to his sons and his X chromosome to his daughters. Mothers transmit one of the other of their X chromosomes to each of their offspring. Therefore, the inheritance of the male's Y chromo-some determines which fetuses will be male (see Figure 1).

The mtDNA, by contrast, is inherited only from the mother. Actually, the mtDNA is the degenerate genome of an ancient symbiotic bacteria that entered the proto-eukaryotic cell about 2 bil-lion years ago. We now call these bacteria mitochondria, and they currently function as the primary energy-generating organelles, or

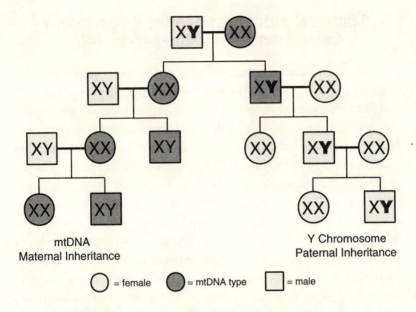

mtDNA
Maternal Inheritance

Y Chromosome
Paternal Inheritance

◯ = female   ⬤ = mtDNA type   ☐ = male

FIGURE 1 The maternal inheritance of the mtDNA is shown along the left lineage and the paternal inheritance of the Y chromosome is shown along the right.
The mother transmits her mtDNA to all of her children, but only her daughters
transmit the mtDNA to the next generation. The father transmits his Y chromosomes to his sons and they transmit the Y to their sons.

power plants, of our cells. Originally, the mitochondria had a
bacterial-sized genome, but over the past 2 billion years most of the
mtDNA genes have been transferred to the nucleus. Today, only 37
genes remain in the human mtDNA: 13 proteins, 2 rRNA genes, and
22 tRNA genes.

The maternal inheritance of the mtDNA is a direct consequence
of its cytoplasmic location in the cell. The oocyte has a large cytoplasm, whereas the sperm has almost none. Consequently at fertili-
zation the oocyte contributes about 200,000 mtDNAs and the sperm
about 100. Moreover, the sperm mitochondria are selectively
destroyed when they enter the oocyte cytoplasm. Therefore, the
female's mitochondria always win, making the mtDNA exclusively
maternally inherited (see Figure 1).

The Y chromosome and the mtDNA have been transmitted sequentially from generation to generation throughout human history, only changing by sequential mutations. Therefore, the number of mutational differences between the Y chromosomes of two individuals is proportional to the time since they shared a common father, and the number of changes in the mtDNA is proportional to the time since they shared a common mother.

Phylogenetic Trees: MtDNA and Y

A phylogenetic tree portrays the number of sequence changes (genetic distance) between any two individual DNA segments. Since sequence divergence is proportional to the time to the most recent common ancestor, Y chromosome or mtDNA comparisons can be used to deduce the genetic relationships between individuals from around the world. The sum of the Y chromosome or mtDNA sequence variants of an individual are designated as that individual's haplotype. A group of related haplotypes is designated a haplogroup. The root of a phylogenetic tree is identified by comparison with a very divergent sequence, known as an outgroup. For trees of human mtDNAs, the outgroup is frequently the chimpanzee (see Figure 2).

The accumulation of neutral mutations over time acts like a molecular clock. For the maternally or paternally inherited mtDNA or Y chromosome, the more mutational differences between two DNAs, the more time since the two individuals shared a common maternal or paternal ancestor.

MtDNA Mutations and the Origins of Women

The accumulation of mtDNA mutations has occurred continuously since humans arose in Africa about 150,000 to 200,000 years before present (YBP) and migrated out of Africa into Eurasia and then to the Americas (see Figure 2). Because certain mutations occurred at critical junctions in human history, they correlate with important times and events in female history. For example, one particular nucleotide change at nucleotide pair 3592 occurred in Africa at the time our

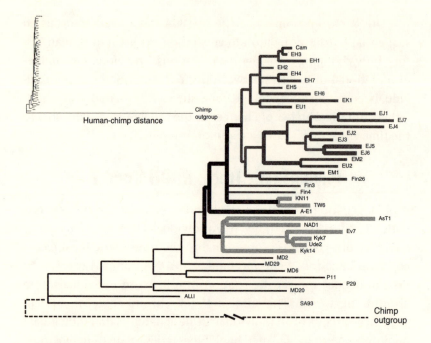

FIGURE 2 Global phylogeny of whole human mtDNA genome sequences. A phylogenic tree shows the amount of DNA sequence changes between comparable pieces of DNA from various individuals. The right tip of each line represents a DNA molecule, and all of its associated changes define its haplotype. Clusters of related haplotypes which share a common ancestor are called a haplogroup. The origin of the human tree is defined by comparison to an outgroup DNA, in this case the Chimpanzee. The relative genetic distance between the human mtDNA sequences and that of the Chimpanzee is shown in the upper left insert. Since the Chimpanzee diverged from human about 5 million YBP, the time required to generate the observed human mtDNA sequence diversity is about 150,000 years.

species arose. Consequently, this variant is found in about two-thirds of all African mtDNAs and defines the ancient macro-haplogroup L. Another variant is found in 40 percent of Europeans and is defined as haplogroup H, whereas four other variants, designated A, B, C, and D, arose in Asia and succeeded in crossing the Bering land bridge to found the Native Americans.

A phylogenetic analysis comparing 39 complete mtDNA sequences from Africa, Asia, and Europe has revealed that all human mtDNAs belong to a single phylogenetic tree and that the mtDNAs from each continent (Africa, Europe, and Asia) cluster together as major branches of the tree. The African mtDNAs are the most divergent, hence the oldest, followed by Asian and then European DNAs. Comparison with chimpanzee mtDNA roots the human mtDNA tree in Africa and gives an estimated coalescence time of about 150,000 YBP (see Figure 2). Hence, our species arose in Africa relatively recently.

While two-thirds of African mtDNAs belong to macro-haplogroup L (encompassing L1 and L2), the remaining African mtDNAs form haplogroup L3, which includes the intermediates between African macro-haplogroup L mtDNAs and those found in Europe and Asia (see Figure 3).

African macro-haplogroup L is subdivided into haplogroups L1 and L2, and within these haplogroups distinct clusters of haplotypes are observed for the !Kung, Western Biaka Pygmies, Eastern Mbuti Pygmies, and Senegalese. Phylogenetic and sequence divergence estimates indicate that the !Kung and Biaka Pygmies are the oldest African populations and that the Mbuti and Biaka Pygmies had independent origins. Hence, in the tropical rain forest the pygmy lifestyle evolved two independent times from two different populations, one early in human history and the other much later. This shows how pliable our physiognomy is; it can change rapidly depending on local environmental conditions.

Using sequence divergence to calculate key events, macro-haplogroup L can be determined to be about 125,000 to 150,000 years old; and the L1 populations of !Kung and Biaka pygmies are nearly as old, indicating that they are representatives of the earliest human populations. Thus, the hunter-gatherer lifestyle of the San Bushmen may be the best approximation today of what it was like to live in Africa at the time of the earliest humans.

Humans expanded throughout Africa, ultimately reaching Ethiopia. While northeast Africa harbors virtually the total spectrum of African mtDNA variations, only two mtDNA lineages, macro-haplogroups M and N, left Africa to colonize Eurasia. Asia was

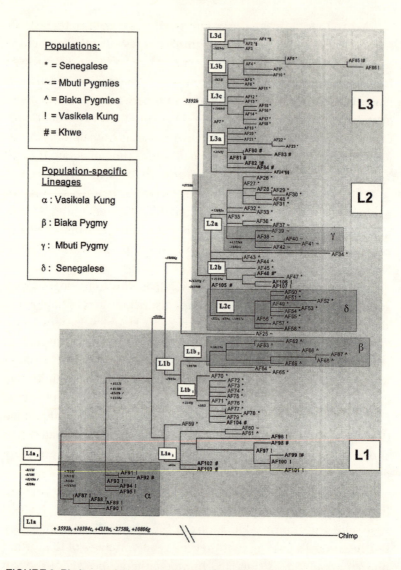

FIGURE 3 Phylogenetic tree of African mtDNA haplotypes. Haplogroups L1 and L2 comprise macro-haplogroup L and are separated from haplogroup L3 by the ancient polymorphism at nucleotide pair 3592. Haplogroup L3 contains the progenitors of the Eurasian mtDNA macro-haplogroups M and N. The mtDNA haplotypes surrounded by boxes α, β, γ and δ were only found in the designated populations. The Biaka Pygmies and the Vasikela Kung are the oldest distinct African populations.

populated about 75,000 years ago from derivatives of both macro-haplogroups M and N (see Figure 4). From these progenitors, multiple Asian-specific lineages arose: N gave rise to haplogroups A, B, F, etc., and M gave rise to C, D, F, G, etc. Furthermore, this lineage diversification occurred regionally. For example, haplogroup F is at its highest frequency in Southeast Asia; B is prevalent along the Asian coast; and A, C, D, and G are concentrated in Siberia.

Similarly, in Europe, where most mtDNA lineages were derived from macro-haplogroup N, nine haplogroups (H, I, J, K, T, U, V, W, and X) account for 98 percent of all mtDNA (see Figure 5). The European mtDNA coalescence time is 40,000 to 50,000 years, and haplogroups U, V, X, and W are among the most ancient lineages.

MtDNA analysis of aboriginal Siberian populations has revealed that haplogroups C and D are widely distributed throughout Siberia, whereas haplogroup A reaches its highest frequency in the Chukchi of the Chukotka peninsula adjacent to Alaska. Haplogroups C, D, and G are also found in the Chukchi. Hence, as humans moved northward, progressively fewer mtDNA types became enriched, until only A, C, and D predominated. As a result, these three mtDNAs crossed the Bering land bridge to give rise to Native Americans (see Figure 6). In addition to A, C, and D, two other mtDNA haplogroups are found in the Americas: B and X. The presence of haplogroup B throughout coastal Asia raises the possibility that it came to the Americas via a coastal migration. (X will be considered later.)

The geographic distribution and subvariation of Siberian and Native American haplogroups A, B, and C also indicate the number of Siberian migrations to the Americas. All North, Central, and South America Paleo-Indian populations encompass varying levels of haplogroups A, B, C, and D. By contrast the Na-Dene-speaking peoples of northwestern North America have only haplogroup A, including distinctive derivatives of the original haplogroup A that gave rise to the Paleo-Indians. Clearly, then, the Paleo-Indian and Na-Dene migrations were separate events. The Paleo-Indian mtDNA variation gives an age of 20,000 to 30,000 YBP, whereas the Na-Dene variation gives an age of 7,000 to 9,000 YBP. Finally, a third

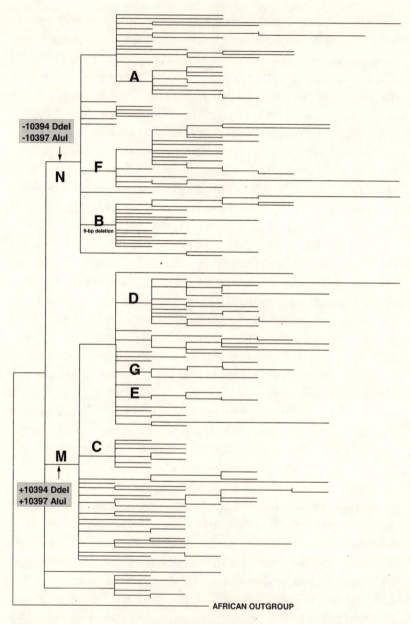

FIGURE 4 Phylogeny of Asian mtDNA haplotypes. All Asian mtDNA are derived from only two macro-haplogroup mtDNA lineages M and N, defined by the presence or absence of a pair of ancient polymorphisms at nucleotide pairs 10394 and 10397.

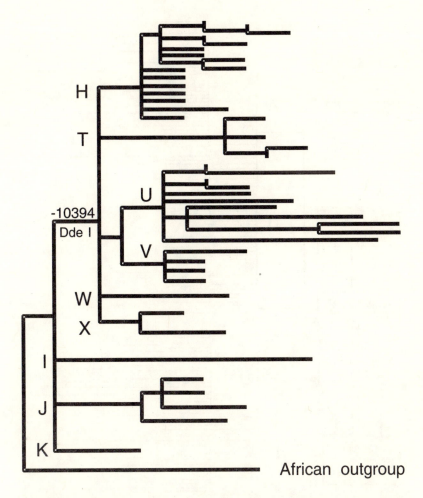

FIGURE 5 Phylogeny of European mtDNA haplotypes. Most European mtDNAs are derived from macro-haplogroup N or its progenitor. The most common European haplogroup is H, representing approximately 40% of European mtDNAs.

expansion out of Chukotka along the Arctic carried haplogroups A and D and gave rise to the Eskimos and Aleuts.

But what about the X? We found haplogroup X when we were studying the Ojibwa of the Great Lakes. When we first encountered this other haplotype, we thought it was due to post- Columbus admixture between Native Americans and immigrant Europeans.

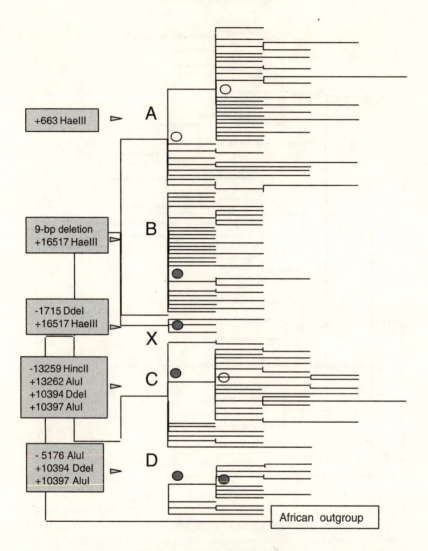

FIGURE 6 Phylogeny of Native American mtDNA haplotypes. All Native American mtDNAs fall into to only five haplogroups. Haplogroups A, B, C, and D are confined to Asia, with haplogroups A, C, and D reaching their highest levels in Siberia. Haplogroup B is found primarily along the Asian coast. Haplogroup X is primarily found in Europe. The dots indicate the Native American founder mtDNAs that are shared between Asians and Native Americans.

However, in Europe H is common and X is rare, but in the Ojibwa X was common and H was rare. Hence, the Ojibwa X was unlikely to have arrived simply by recent European input. To examine this question further, we compared the sequence variation between European haplogroup X and Native American haplogroup X mtDNAs (see Figure 7). This revealed that the Native American haplogroup X mtDNAs are totally distinct from those of the Europeans. Moreover, we were able to calculate that the Native American and European haplogroup

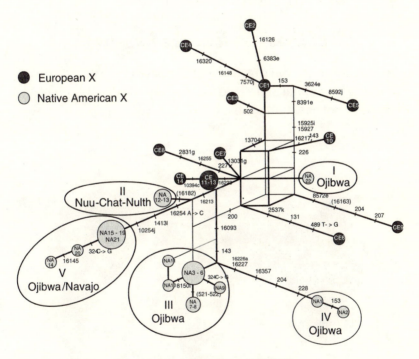

European - Native American divergence time ~15,000 YBP

FIGURE 7 Phylogenetic network of European and Native American haplogroup X mtDNA haplotypes. European mtDNA haplotypes are shown as filled circles, while the Native American haplotypes are shown as open circles. None of the Native American haplotype X mtDNAs have the same sequence motif as the European mtDNAs demonstrating that the Native American haplogroup X mtDNAs are not the result of recent European mixing with Native American populations.

X mtDNAs shared a common ancestor, about 15,000 YBP. Hence, the Native American haplogroup X arrived in the Americas long before Columbus and must represent yet another ancient migration to the New World originating from either Asia or Europe.

In summary, our studies on human mtDNA variation have confirmed that modern humans came out of Africa to occupy all of the continents of the world within the past 100,000 years (see Figure 8). Following colonization of Asia and Europe, a limited number of individuals succeeded in surviving the subarctic conditions of Siberia and Beringia and colonized the Americas.

What About the Men, the Forgotten Y?

The study of mtDNA has provided striking insights into the history of our female ancestors all the way back to the proverbial Eve. But what about Adam and our male ancestors?

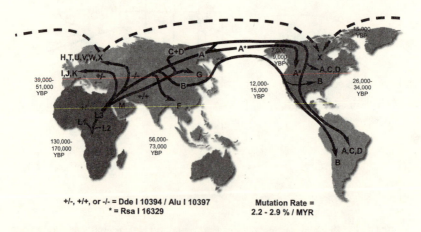

+/-, +/+, or -/- = Dde I 10394 / Alu I 10397
* = Rsa I 16329

Mutation Rate =
2.2 - 2.9 % / MYR

FIGURE 8 Global migrations of women as defined by mtDNA variation. Haplogroup names and their continent of origin are indicated by capital letters. The 10394/10397 polymorphisms defining macro-haplogroups M and N are shown by the +/- symbols. The approximate ages of each migration are shown. The dotted line represents the possible routes by which haplogroup X could have migrated from Europe to north-central North American about 15,000 YBP.

The Y chromosome, a large linear molecule of about 60 megabases, provides similar insights into our patrilineal history. Like the mtDNA phylogeny, the Y chromosome phylogenetic tree is rooted in Africa and has continent-specific branches. Generally, the Y chromosome results parallel the mtDNA results, which is a relief.

Studies of Siberian and Native American Y chromosomes demonstrate how beautifully the Y chromosome data can compliment the mtDNA data. The distribution and frequency of Y chromosome variants have revealed two major migrations from Siberia to the Americas (see Figure 9). The first arose in central Siberia with Y chromosome haplogroup M45a. As this lineage migrated northeast into Chuktoka, it gave rise to the new lineage, M3. M45a and M3

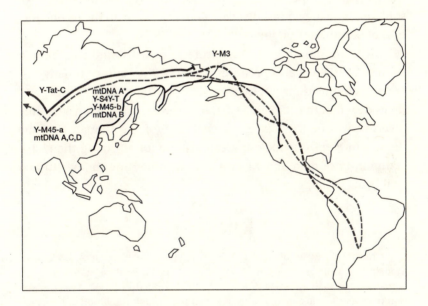

FIGURE 9 Correlation of the mtDNA (female) and Y chromosome (male) migration patterns from Siberia to the Americas. Two independent migrations are observed. The first originated in central Siberia, crossed through Chukotka and colonized all of North, Central and South America giving rise to the Paleo-Indians. The second originated in eastern Siberia from the Sea of Okhotsk/Amur River region, passed through Chukotka and occupied northwestern North America to found the Na-Dene populations.

then expanded into North, Central, and South America along with the mtDNA lineages A, C, and D to give the Paleo-Indians. The second migration arose in eastern Siberia along the Okhotsk with the Y chromosome lineages M45b and S4Y-T. These lineages crossed the Bering land bridge along with the modified mtDNA A to form the Na-Dene (see Figure 9). Thus, the mtDNA and the Y chromosome tell similar stories.

Conclusion

Studies of the genomic diversity of the human mtDNA and Y chromosome confirm the recent African origin of our species and provide a strikingly detailed description of the subsequent colonization of Asia, Europe, and the Americas. However, the importance of studying human genome diversity does not end with the reconstruction of our ancient origins.

An understanding of the nature and distribution of human genomic variation also promises to reveal the causes of many diseases, including diabetes, hypertension, cardiovascular disease, and neurodegenerative disease. Characterization of human genomic diversity will be the next great challenge of the Human Genome Project and will provide our greatest hope for affirming the global community of men and women and of promoting the health and well-being of all peoples.

References

Cruciani, F., P. Santolamazza, P. Shen, V. Macaulay, P. Moral, A. Olckers, D. Modiano, S. Holmes, G. Destro-Bisol, V. Coia, D. C. Wallace, P. J. Oefner, A. Torroni, L. L. Cavalli-Sforza, R. Scozzari, and P. A. Underhill. 2002. A back migration from Asia to sub-Saharan Africa is supported by high-resolution of human Y-chromosome haplotypes. *American Journal of Human Genetics* 70:1197-1214.

Lell, J. T., R. I. Sukernik, Y. B. Starikovskaya, B. Su, L. Jin, T. G. Schurr, P. A. Underhill, and D. C. Wallace. 2002. The dual origin and Siberian affinities of Native American chromosomes. *American Journal of Human Genetics* 70:192-206.

Wallace, D.C., M. D. Brown, and M.T. Lott. 1999. Mitochondrial DNA variation in human evolution and disease. *Gene* 238:211-230.

Daniel J. Kevles

Eugenics, the Genome, and Human Rights

Several years ago President Clinton remarked that the next half-century will be the age of biology and that the engine of that age will be genetics. Partly because of the Human Genome Project, scientists are producing a torrent of information and claims about the role of genes in human disease, capacities, and behavior. The new knowledge is bringing about a revolution in the diagnosis of diseases and disorders. It is also predicted to yield a powerful arsenal of therapies and cures—and possibly an ability to improve people genetically. Indeed, some fear that it threatens to fulfill the long-standing dream of the eugenics movement that flourished early in the twentieth century and trampled on human rights.

Eugenics has no more powerful association than with the Nazis. In Germany during the Hitler years, the eugenics movement prompted the sterilization of several hundred thousand people and helped lead to anti-Semitic programs of euthanasia and ultimately, of course, death camps. The association of eugenics with the Nazis is so strong that many people were surprised at the news several years ago that between the 1930s and the 1970s Sweden had sterilized some 60,000 people, most of them women, initially with the intention of reducing the births of children suffering from genetic diseases and disorders. The fact of the matter is that after the turn of the century, eugenics movements, including demands for sterilization of the unfit, blossomed in the United States, Canada, Britain, and Scandinavia, not to mention elsewhere in Continental Europe and parts of Latin America and Asia. Eugenics was thus not unique to the Nazis. It could—and did—happen everywhere.

Modern eugenics was rooted in the social Darwinism of the late nineteenth century, with all its metaphors of fitness, competition, and rationalizations of inequality. Indeed, the word "eugenics" was coined by Francis Galton, a cousin of Charles Darwin and an accomplished scientist in his own right. He promoted the ideal of improving the human race by, as he put it, getting rid of the "undesirables" and multiplying the "desirables." Eugenics began to flourish after the rediscovery in 1900 of Mendel's theory that the biological makeup of organisms is determined by certain "factors," later identified with genes. The application of Mendelism to human beings reinforced the idea that we are determined almost entirely by our "germ plasm."

Eugenics was by no means a crackpot movement. Its doctrines were articulated by physicians, mental health professionals, and scientists, notably biologists, who were pursuing the new discipline of genetics, and medical practitioners, especially those who worked with people suffering from mental diseases and disorders. The doctrines were widely popularized in books, lectures, and articles to the educated public of the day, and they were bolstered by research that poured out of institutes for the study of eugenics or "race biology" that were established in a number of countries, including Denmark, Sweden, Britain, and the United States. A chart displayed at the Kansas Free Fair in 1929 was designed to illustrate the "laws" of Mendelian inheritance in human beings, stating that "unfit human traits such as feeblemindedness, epilepsy, criminality, insanity, alcoholism, pauperism, and many others run in families and are inherited in exactly the same way as color in guinea pigs."

The experts raised the specter of social "degeneration," insisting that "feebleminded" people—to use the broad-brush term then commonly applied to persons believed to be mentally retarded—were responsible for a wide range of social problems and were proliferating at a rate that threatened social resources and stability. Feebleminded women were held to be driven by a heedless sexuality, the product of biologically grounded flaws in their moral character that led them to prostitution and illegitimacy. Hereditarian biology attributed poverty and criminality to bad genes rather than flaws in the social corpus.

Although frequently assumed to have been essentially a socially conservative movement, eugenics in fact belonged in no small part to the wave of progressive social reform that swept through Western Europe and North America during the early decades of the twentieth century. For progressives, eugenics was a branch of the drive for social improvement or perfection that many reformers of the day thought might be achieved through the deployment of science to good social ends. Eugenics, of course, also drew significant support from social conservatives, concerned to prevent the proliferation of lower-income groups and save on the cost of caring for them. The progressives and the conservatives found common ground in attributing phenomena such as crime, slums, prostitution, and alcoholism primarily to biology and in believing that biology might be used to eliminate these discordances of modern urban industrial society.

Race was a minor subtext in Scandinavian and British eugenics, but it played a major part in the American and Canadian versions of the creed. North American eugenicists were particularly disturbed by the immigrants from Eastern and Southern Europe who had been flooding into their countries since the late nineteenth century. They took them to be not only racially different from but also inferior to the Anglo-Saxon majority, partly because they were disproportionately represented among the criminals, prostitutes, slum dwellers, and feebleminded in many cities. Anglo-American eugenicists fastened on British data that indicated that half of each succeeding generation was produced by no more than a quarter of its married predecessor and that the prolific quarter was disproportionately located among the dregs of society. Eugenic reasoning in the United States had it that if immigrant deficiencies were hereditary and Eastern European immigrants outreproduced natives of Anglo stock, the quality of the American population would inevitably decline.

Eugenicists on both sides of the Atlantic argued for a two-pronged program that would increase the frequency of socially good genes in the population and decrease that of bad genes. One prong comprised "positive" eugenics, which meant manipulating human heredity and/or breeding to produce superior people. The other was "negative" eugenics, which meant improving the quality of the human race by eliminating or excluding biologically inferior people from the population.

In Britain between the world wars, positive eugenic thinking led to proposals—they were unsuccessful—for family allowances that would be proportional to income. In the United States it fostered so-called Fitter Family competitions, a standard feature at a number of state fairs that were held in their "human stock" sections. At the 1924 Kansas Free Fair, winning families in the three categories—small, average, and large—were awarded a Governor's Fitter Family Trophy. "Grade A Individuals" received a medal that portrayed two diaphanously garbed parents, their arms outstretched toward their (presumably) eugenically meritorious infant. It is hard to know what made these families and individuals stand out as fit, but some evidence is supplied by the fact that all entrants had to take an IQ test—and the Wasserman test for syphilis.

Much more was urged for negative eugenics, notably the passage of eugenic sterilization laws. In the United States by the late 1920s, such laws had been enacted in two dozen American states, largely in the Middle Atlantic region, the Midwest, and California, the champion. As of 1933, California had subjected more people to eugenic sterilization than had all other states of the union combined. Similar measures were passed in Canada in the provinces of British Columbia and Alberta. Almost everywhere they were passed the laws reached only to the inmates of state institutions for the mentally handicapped or mentally ill. People in private care or in the care of their families eluded them. They thus tended to work discriminatorily against lower-income and minority groups. California, for example, sterilized blacks and foreign immigrants at nearly twice the rate as their presence in the general population.

The sterilization laws implicitly rode roughshod over private human rights, holding them subordinate to an allegedly greater public good. Such reasoning figured explicitly in the U.S. Supreme Court's 8-to-1 decision, in 1927, in the case of *Buck v. Bell*, which upheld Virginia's eugenic sterilization law. Justice Oliver Wendell Holmes, writing for the majority, averred:

> We have seen more than once that the public welfare may call upon the best citizens for their lives. It would be strange if it could not call upon those who already sap the strength of the State for these lesser sacrifices, often not felt to be such by those concerned, in order to prevent our being swamped with incompetence. It is better for all the world, if instead of waiting to execute degenerate offspring for crime, or to let them starve for their imbecility, society can prevent those who are manifestly unfit from continuing their kind. The principle that sustains compulsory vaccination is broad enough to cover cutting the Fallopian tubes. . . . Three generations of imbeciles are enough.

In Alberta the premier called sterilization far more effective than segregation and, perhaps taking a leaf from Holmes's book, insisted that "the argument of freedom or right of the individual can no longer hold good where the welfare of the state and society is concerned."

Sterilization rates climbed with the onset of the worldwide economic depression in 1929. In parts of Canada and the Deep South and throughout Scandinavia, it acquired broad support, not primarily on eugenic grounds (though some hereditarian-minded mental health professionals continued to urge it for that purpose) but on economic ones, raising the prospect of reducing the cost of institutional care and poor relief. Madge Thurlow Macklin, a geneticist at the University of Western Ontario, an organizer of the Eugenics Society of Canada and an outspoken advocate of eugenic sterilization of the feebleminded, warned against the differential birth rate, declaring: "We care for the mentally deficient by means of taxes, which have to be paid for by the mentally efficient. . . ." Even geneticists who disparaged sterilization as a panacea against degeneration held that sterilization of the mentally disabled would yield a social benefit because it would prevent children being born to parents who could not care for them.

In this intensified drive for sterilization, individual human rights were once again held to be subordinate to some greater social good—but especially in this era to some greater economic good. In Scandinavia, sterilization was broadly endorsed by Social Democrats as part of the scientifically oriented planning of the new welfare state. Alva Myrdal spoke for her husband, Gunnar, and for numerous liberals like themselves in 1941 when she wrote: "In our day of highly accelerated social reforms the need for sterilization on social grounds gains new momentum. Generous social reforms may facilitate home-making and childbearing more than before among the groups of less desirable as well as more desirable parents. . . . [Such a trend] demands some corresponding corrective." On such foundations, among others, sterilization programs continued in several American states and Alberta as well as Scandinavia well into the 1970s.

However, during the interwar years, eugenic doctrines were increasingly criticized on scientific grounds and for their class and racial bias. It was shown that many mental disabilities have nothing to do with genes, that those which do are often complicated rather than simple products of them, and that most human behaviors, including the deviant variety, are shaped by environment at least as much as by biological heredity, if they are fashioned by genes at all. Science aside, eugenics became malodorous precisely because of its connection to Hitler's regime, especially after World War II, when its complicity in the Nazi death camps was revealed.

All along many people on both sides of the Atlantic had ethical reservations about sterilization and were squeamish about forcibly subjecting people to the knife. Attempts to authorize eugenic sterilization in Britain had reached their high-water mark in the debates over the Mental Deficiency Act in 1913; they failed not least because of powerful objections from civil libertarians insistent on defending individual human rights. More than a third of the American states declined to pass sterilization laws, and so did the eastern provinces of Canada. Most of the American states that did pass such laws declined to enforce them, and British Columbia's law was enforced very little.

The opposition comprised coalitions that varied in composition, drawing from scientifically dubious mental health professionals and civil libertarians, some of whom warned that compulsory sterilization constituted a Hitler-like suppression of private reproductive rights. In Alabama, for example, attempts to pass a sterilization law in the mid-1930s prompted a Methodist newspaper to warn that the "proposed sterilization bill is a step" toward the "totalitarianism in Germany today." There, it was said, the "state is taking private matters—matters of individual conscience, and matters of family control—in hand, and sometimes it's a rough hand, and always it's a strong hand." Governor Bibb Graves put the issue more succinctly: "The great rank and file of the country people of Alabama do not want this law; they do not want Alabama, as they term it, Hitlerized."

Sterilization was also vigorously resisted by Roman Catholics, partly because it was contrary to Church doctrine and partly because a very high fraction of recent immigrants to the United States were Catholics and thus disproportionately placed in jeopardy of the knife. For many people before World War II, individual human rights mattered far more than those sanctioned by the era's science, law, and perception of social needs.

The revelations of the Holocaust strengthened the moral objections to eugenics and sterilization, and so did the increasing worldwide discussion of human rights, a foundation for which was the Universal Declaration of Human Rights that the United Nations General Assembly adopted and proclaimed in 1948. Since then the movement for women's rights and reproductive freedom has further transformed moral sensibilities about eugenics so that today we recoil at the majority's ruling in *Buck v. Bell*.

Let's return now to our own day and the Human Genome Project. At the moment, the social and ethical challenges arising from molecular genetics do not appear to lie in a recrudescence of eugenics. Rather they center in considerable part on the grit of what science is producing in abundance: genetic information. They center on the control, diffusion, and use of that information in the context of a socially charged market economy.

Much of the discussion about the information to come from the Human Genome Project has rightly emphasized issues of individual human rights—that employers may seek to deny jobs to applicants with a susceptibility or an alleged susceptibility to disorders such as manic depression or illnesses arising from features of the workplace. Life and medical insurance companies may well wish to know the genomic signatures of their clients, their risk profiles for disease and death. In the public realm, as the costs of medical care continue to rise, the increasing acquisition of genetic information could conceivably lead to a renewal of the ethical premises of the original eugenics movement, an insistence that the reproductive rights of individuals must give way to the medical-economic welfare of the community as a whole.

To be sure, the likelihood of a new eugenics movement seems small. Resistance to such ventures is high across American society. The Catholic Church remains adamantly opposed, and vulnerable groups—particularly racial minorities and people with handicaps—are far more empowered now than they were in the early 1900s. Legal and constitutional rights of privacy and reproduction comprise a strong bulwark against the revival of any state-mandated eugenics programs. In our day the principle that sustains compulsory vaccination is no longer broad enough to cover cutting the Fallopian tubes or any other state interference with reproduction. Even so, the record of eugenics in North America and Northern Europe offers a powerful indication that the uses of genetic science and genetic information today warrant considerable care and attention not only in law but also in practice to civil liberties, individual rights, and social decency. History has taught us at the least that concern for individual rights belongs at the heart of whatever stratagems we may devise for deploying our rapidly growing knowledge of human and medical genetics.

David J. Rothman
Sheila M. Rothman

Redesigning the Self

The Promise and Perils of
Genetic Enhancement

 The term *genetic enhancement* is used to describe efforts to make individuals better than well, optimizing their capabilities by taking them from standard levels of performance to peak performance. This raises some intriguing questions because rather than make a copy of an individual, which is what would happen with cloning, genetic enhancement may be able to improve that individual, which might be more appealing. And without being too egocentric, it would seem to me that, socially speaking, a procedure or technology that could enhance us and not just our children would be all the more attractive.

Although the distinction between cure and enhancement has a surface logic, it has surprisingly little meaning in establishing a biomedical research agenda, in dictating medical practice, or in formulating health policy. To the contrary, cure and enhancement merge into each other and actually feed off each other, with interventions that begin in an effort to cure often quickly becoming enhancements.

What will the new genetic enhancement technology bring? Although it is too soon to be certain, there is good reason to think that in the next 10 to 20 years we will have developed genetic enhancements to improve memory and perhaps problem-solving ability; to reduce dramatically the level of and need for sleep; to improve physical capacities to make us stronger and quicker; to provide perfect pitch; to provide personality traits, including higher levels of aggression or perhaps higher levels of altruism; to improve immunology and protections against diseases, such as cardiac disease and cancer; and to provide protections against weight gain and for increased longevity.

Concerns About Enhancement

In discussions about enhancement there is a kind of diffuse opposition to genetic enhancement, or at least anxiety about it and a certain amount of unease, for several reasons. First, there seems to be something unnatural about these efforts. That is, one should not be tinkering with nature—altering genetic codes for enhancement seems to violate the human condition. The literary world launched us into these discussions with the publication of, for example, Shelley's *Frankenstein*. Perhaps the anxiety is most clearly expressed in and around the possibility of longevity. A society in which everyone lives to be 200 or 250 years old arouses our instincts that there is something unnatural, grotesque, and maybe even unwanted about such a world.

The second source of anxiety is of a somewhat different nature. Genetic enhancement may strike some as being frivolous. It will be linked in the popular mind to having cosmetic surgery, making excuses to eat cheeseburgers while not getting fat and elevating our cholesterol levels, changing the color of our eyes or hair or even our skin. These objections are serious and are based in part on the belief that other goals should be given scientific priority.

There is also anxiety about the possible misuses of genetic enhancements, with the obvious specter of Nazi Germany in the background. Enhancement is linked to eugenics and the fear that in the hands of the state it will become a tool of oppression—not only gross oppression but also milder forms of coercion. For example, the hearing-impaired community believes that genetic enhancement might be defined in such a way that a fetus determined to be deaf would be enhanced by taking away the deafness. They would not define this as cure because they do not think of themselves as sick or deviant, but they would see the removal of a hearing disability from a fetus as an enhancement and a state intrusion.

Finally, the fear is often raised that genetic enhancement technologies will be monopolized by the well-to-do at the expense of others and will widen the gap between classes, giving the rich still more advantages. Thus, it will be the "haves" that will become enhanced, which will provide them with a biological edge in addition to their existing economic edge. These objections, however, no matter how much you may empathize with them, will not slow the drive to enhancement. In the end the engine driving enhancement runs on many cylinders, including science, medicine, commerce, and culture, and it will brook no interference. In addition, the real problems may not rest in the issues that have already been identified, but in another arena—that of risk. The problem may well be an impatience, an unbridled enthusiasm, and in the end it is possible that the drive to enhancement will generate more harm than good.

The Concept of the Natural

The concept of the natural does not deter science. In fact, I would put it the other way. To do battle with the natural seems to be a critical driving force. Science has a long and deep history of completely disrespecting the concept of the natural. Well through the nineteenth and twentieth centuries, this scientific attitude affirmed the right and the need to know and a rejection out of hand of the concept of forbidden knowledge, or really even forbidden tinkering. This mindset celebrates the pursuit of nature in order to learn her secrets and strip her of them.

To illustrate the deep sense of just how ingrained these attitudes are within the scientific mindset, I start with Claude Bernard, known to many as the father of experimental science and physiology and discoverer of glycogen. He wrote: "Man becomes an inventor of phenomena, a real foreman of creation. And under this head we cannot set limits to the power that he may gain over nature through future progress." Experimental sciences, as Bernard defined them, were active sciences. The goal Bernard set out for physiology in the mid-nineteenth century is "to conquer living nature, act upon vital phenomena, regulate them and modify them." Sensing that there might be objections to this approach, Bernard told his fellow scientists that they should pay no attention to potential objections from outside science: "It is impossible for men judging facts by such different ideas ever to agree. A man of science should attend only to the opinion of men of science who understand him,"—a fabulous, perhaps even terrifying, line.

A reading of Charles Darwin in the late nineteenth and early twentieth centuries also gave men of medicine and science the right to actively contravene the conventional bounds of nature and the natural. In the Darwinian world, after all, change over time was inevitable, but the source of that change, as Darwin outlined it in *The Origin of Species*, was natural but happenstance—accidental. The results of natural selection were the nonhuman and unwilled processes of selection. In fact, this position inspired many scientists to say that if evolution works through a haphazard method, we should take this as a challenge. Why leave it to the haphazard?

The well-known biologist Jacques Loeb (who was fictionalized as Max Gottlieb in Sinclair Lewis's *Arrowsmith*) articulated in the early 1900s the notion that the natural order is nothing more than the result of chance mutation and that chance should become subsidiary to our ability to affect the natural. In other words, mutations are accidental, so why wait for the accident? He argued that science, not chance, should create the variation. Human design is at least as desirable as chance in altering nature. Loeb expressed this quite clearly, saying that investigators should find a variety of ways "for the transformation of the species beyond that which we have at present." The scientist could do better than nature. The scientist, he fantasized, might actually be able to halt the wasting of the body in old age.

Loeb was not alone in this thinking. In 1923 the biochemist J. B. S. Haldane wrote a best-selling pamphlet in England called *Daedalus, or Science and the Future*. Haldane's premise was that the idea of limiting science on the grounds that what was should determine what should be had to be rejected. "There is no great invention," he wrote, "from fire to flying that has not been hailed as an insult to some God." He looked to the abolition of disease and was happy in promoting what he called, "the direct improvement of the individual" or what we today call enhancement. Eager to do it through medicine, Haldane used endocrinology as one of his cases in point. "As our knowledge of this subject increases, we may be able, for example, to control our passions by some more direct method than fasting and flagellation, to stimulate our imagination by some reagent with less after-effects than alcohol, to deal with perverted instincts by physiology rather than prison."

Anticipating the identification and synthesis of estrogen, he speculated that as we learned more about the chemical substances produced by ovaries and were able to isolate and replicate them "we shall be able to prolong a woman's youth and allow her to age as gradually as the average man."

The expressions I have listed from science found popular reception. Aldous Huxley's *Brave New World* picks up on one of Haldane's fantasies about "test-tube children." However, as Haldane concedes in an aside: "Man armed with science is like the baby with a box of matches." But he concluded: "It is science that will enable man to refashion his own body and those of other living beings, so as to be able to overcome the dark and evil elements in his own soul."

H. G. Wells picked up these themes in his 1895 essay, *The Limits of Individual Plasticity*. Wells advocated no mere subservience to natural selection, stating that we are raw material that should be reshaped and altered. He expressed a genuine sense that the role of science is not to be disrespectful of where we are but to use that knowledge to take us to new areas. Respect for the natural will only incite science, not limit it.

The Argument Against Frivolity

As to the frivolity of enhancement, this will not prove to be very much of a barrier. The real problem is that what any one of us might consider frivolous is an absolutely critical intervention to somebody else. Frivolity is in the eye of the beholder. An example can be found in the use of growth hormone, known since the 1950s as an extract from the pituitary gland that was later synthesized and administered initially to those with growth hormone deficiencies to increase eventual height. But then the question arises of whether it should be used as an enhancement, that is, for short children who are not growth hormone deficient.

When some bioethicists address the matter, they frame it in terms of the athletic father who wants his short son to grow up to play for the New York Knicks. Framed this way, the issue of frivolity is clear. Such a use of synthetic growth hormone is frivolous; it should not be allowed. But the framing of the question is wrong. The issue is not whether my child should grow up to play for the Knicks but rather what I can do for my 12-year-old girl who is 4 feet, 5 inches tall and has to shop in the children's section while all of her friends go to the teen or adult sections of the clothing store. Or what I do for my young son who is 4 foot, 8 inches tall, the mascot of his school, and the butt of cruel jokes. To these parents and to the children as patients, this is not a frivolous matter. It is about trying to make their way in the world with less pain. And I emphasize this even at the risk of overdramatizing it, because when such a child and parent enter the physician's office, the physician may well respond to this very unhappy child and distraught parent. The physician may be prepared to use growth hormone as an enhancement to genuinely attempt to bring some happiness to troubled lives.

But this happiness leaves a door very much ajar. Whether it is growth hormone for an adolescent or cosmetic surgery for a young woman, medicine often finds it appropriate to intervene to alleviate unhappiness, or at least to try to reduce it by a degree or two. Can this be abused? Without any question. But there is no denying that the impulse to treat is an impulse that will not stop at the door of genetic technologies.

Will the history of the twentieth century give us pause when it comes to genetic enhancements? The Nazi experience will not be forgotten, but its relevance to these issues may be fairly distant. It is not the heavy hand of the state that evokes concerns today, as it was in Nazi Germany, but the power of attraction to these possibilities on the part of scientists, physicians, and prospective patients. They eagerly, even desperately, seek to create or receive such interventions. Enhancement technologies will not be coerced but embraced.

Justice

Should genetic enhancement be restricted by claims of distributive justice? It is an appealing argument, but it holds very little weight in the sense that we simply do not limit innovation by such standards. Why put genetic enhancement to that test? We are not going to limit access to new imaging machines, and we are not going to limit access to the Internet, although we know there will be unequal distribution of such access. In fact, one could argue that here there may well be a social drive to spread enhancement technologies more broadly. It would be to everybody's interest to raise physical capacity, mental acuity, and disease resistance. Will it work that way? One cannot be sure. But I do not think that invoking distributive justice is sufficient to put much of a crimp in the drive for enhancement.

Who Will Calculate the Risks?

So why not embrace enhancement with wholesale enthusiasm, cast caution to the wind, and take up the agenda freely? As much as we should be trying to conquer cancer, let us try to move forward on enhancement for one very critical reason: innovation cannot be separated from calculations of risk. When we move down the new road, there will be risks and a very clear risk-benefit calculus to be made, and in this area of enhancement such calculations may not get their due.

There are two major reasons why I think risks may well get buried. One is that the process of innovation will make it appear as though enhancement technologies are safe and therefore obviate the need to make a risk calculus. There is good reason to believe that enhancement technologies will rise, not as the result of enhancement research but as an aside or offshoot of research directed at fighting a variety of diseases.

The model might well be Alzheimer's disease. As we learn more about memory loss, it may well be expected that we will learn more about enhancing memory. And technologies devised in the first instance to conquer disease may well spread over quickly and without a different risk calculus into the terms of enhancement. Another example of this is provided by the use of estrogen replacement therapy, which began as a way to help women cope with deeply troubling symptoms as they entered menopause and which is now being widely prescribed to women postmenopausally regardless of their symptoms. Still another example is the trend toward the use of serotonin reuptake inhibitors not only by people with clinical depression but also by those who are unhappy. In each of these instances, that which begins as therapy becomes enhancement without a new risk-benefit calculus. Another reason that risks will be minimized is that it is difficult to identify participants in the health arena who will take risk as their mandate.

It certainly will not be the pharmaceutical or biotechnology industries that will do the risk calculus. Physicians, in general, do not do a very good job in everyday practice of alerting patients to risk. Moreover, the medical profession is more likely to be battling over who gets the enhancement turf than worrying about the risk. The federal government is unlikely to step in because of avoidance of the off-label use of approved interventions. Scandals may evoke a response, but it is unlikely that the government will become the fundamental protectors against risk.

Will patients do the calculations? This is a difficult proposition because peak performance is so valued in our culture. Imagine what we could do with three to five more hours a day to write, trade, or maneuver. Ours is a society that rewards those who have an edge, and if medical technology will provide it, there will be ready consumers. We will rationalize the risks by saying that risks are something that are applied to others, not us.

What Will the Future Bring?

Scientists will continue with talent and impatience to war with nature. Physicians will readily adopt any and all clinical uses of their products. The pharmaceutical and biotechnology industries will do all in their powers to bring their products to physicians' offices and directly to consumers through advertising. And consumers will embrace enhancement with enthusiasm.

What will come of all this? One cannot say with certainty. One scenario might be an increased awareness of risk and calls to regulate it. But I fail to see the constituencies that will make this happen. I believe we are in store for a very different future. Each individual will make his or her own private calculation of risks and benefits without much guidance on how to make the choice. There will be winners, and there will be losers. The problem is that we do not yet know whether the winners will be the risk takers or the risk averse.

Part IV

Financial, Legal, and Ethical Issues and the New Genomics

Michael Waldholz

Introduction

The elucidation and sequencing of the human genome is fraught with great promise and peril. As
Americans living in a democracy and a capitalist
society—unless we have been truly ignoring the
world humming about us—we are cognizant of the
extraordinary role that commerce, business ventures, venture capital, the stock market, entrepreneurs, and entrepreneurial scientists
have played over the past few decades in exploiting and promoting
the genomic revolution. These players have been the driving engines behind the explosion in research into the human genome.

In 1990 I co-authored a book called *Genome* (New York: Simon
and Schuster) with Jerry Bishop, my longtime colleague at the *Wall
Street Journal*. That book was published before James Watson helped
launch the federally funded Human Genome Project. From our perch
at the *Wall Street Journal* back then, covering the doings of the giant

pharmaceutical and burgeoning biotechnology businesses, we could see that research into the genetic basis of sickness and health was becoming the treasure trove of information upon which these businesses would grow.

Despite this great promise, the genome was not sequenced in a day, which has granted us time to consider the implications of this knowledge. In June 2000, Craig Venter and Francis Collins met at the White House to celebrate the sequencing of the 3 billion letters that make up DNA, culminating years of work by federally and privately funded efforts. Over that 10-year period, the tools and information from the genome sequencing have been filtering into the hands of drug, vaccine, and diagnostic developers. In recent years the volume of information has begun to overwhelm industry. In response, legions of start-up companies have been born, racing to find new ways to use this information to beat cancer and fight heart disease, arthritis, diabetes, asthma, and a host of other common ailments that continue to defy adequate treatments.

These companies are scouring the genome for those sets of genes that in altered or varied form lead to disease. We have come to learn that cancer is a genetic disease; that is, it is not necessarily inherited, but the engine that transforms a healthy cell into a tumor is a genetic one. It now appears that most every ailment follows this model—almost every common disease, from heart disease to cancer, has a significant genetic basis, which is not to discount the critical roles of diet and environment in the development of those illnesses. But by unlocking the genetic secrets of disease a huge industry of drug hunters hope to arrive at wholly new ways of identifying and treating disease.

All of this effort, as in any inherently commercial venture, leads to many types of conflicts, including financial, legal, and ethical concerns. There is plenty of money to be made. There are new companies arising, stock prices exploding, intellectual property issues being debated, conflicts of interest to contend with, a few questionable clinical trials to address, and ethical concerns about the use and possible misuse of genetic information. This convergence of events should encourage us to pursue the promises of genomics without throwing caution to the wind.

Kris H. Jenner

Investing in the Biotechnology Sector

 I strongly believe that for years to come investing in the biotechnology sector will be extremely attractive. Numerous biotechnology companies, many of which exist today, will bring therapies to the marketplace that will minimize and potentially cure the pain and suffering of human illness. Three secular trends will dominate the investment landscape in biotechnology for the next 5 to 10 years.

First, we are currently in the "golden era" of biology, with the completion of the Human Genome Project analogous to the earlier completion of the periodic chart of elements. During this era, biological discovery is accelerating, resulting in the development of safer and more efficacious drugs. The insights of basic science usually lead to the creation of new wealth, and the science of the human genome is a marvelous example of this. However, we must keep our feet on the ground because investing in this sector involves a great deal of risk. To develop a drug costs $250 million to $300 million, and failure rates are high for any drug under consideration. It also takes many years to move a drug from a scientific idea to a form approved by the Food and Drug Administration, and the role government plays in this area may increase.

What does the golden era of biology mean for us? Looking at the top drugs sold today, a vast majority of them come from four different types of protein classes (see Box 1). The Human Genome Project offers a substantial expansion of the number of potential targets (historically about 500 different potential targets) for proteins that could lead to the development of safer and more efficacious drugs (see Figure 1). New sciences, including genomics, proteomics, high-throughput screening, and information technologies, are being combined to produce more robust pipelines with lower failure rates and to aid in the development of better medicines for the future.

The second trend is that there is a tremendous need to improve therapies for chronic and acute illnesses. We probably all know

---

Box 1
Golden era of biology

Eighty percent of top 100 prescription drugs target four classes of proteins

Number of Drugs	Protein Class
36	Enzymes
22	GPC receptors
12	Ion channels
9	Nuclear hormone receptors

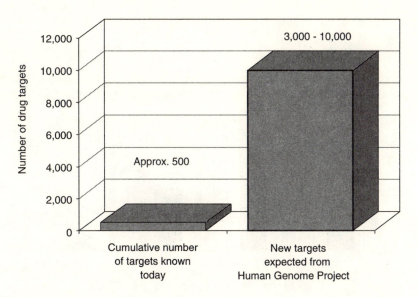

FIGURE 1 Drug innovations expected from the Human Genome Project.

someone who has suffered from an illness for which there is no appropriate or useful therapy. There is a world full of diseases for which better therapies are needed, and one can determine the medical need for improved therapies by looking at the number of people affected by a particular disease.

The third trend, which will dominate the investment landscape for some time to come, involves demographics. The cohort of the population that is most rapidly growing is also the one that is the biggest consumer of pharmaceuticals. In particular, Baby Boomers are the most rapidly growing part of our population, and they will also be the ones who will use the most medications (Figures 2 and 3).

Pharmaceutical Development: Some Realities

Although these three trends will drive the industry, it will still take a considerable amount of time to develop new pharmaceuticals. Part of the promise of some biotechnology companies today is reduction

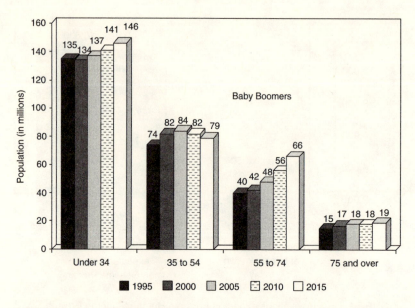

FIGURE 2 U.S. population by age cohorts in five year intervals, 1995-2015. Demographics favor greater demand.

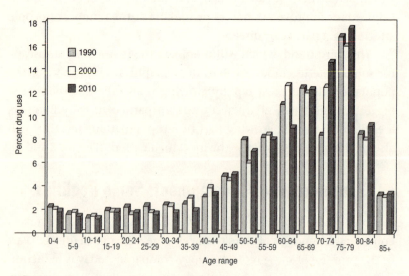

FIGURE 3 Drug use by age in the United States.

in the time required to get drugs to the marketplace, but it will still take many years to get a new product on the market.

During the early phase of development, there is a large consumption of cash, but the successful companies experience an explosive takeoff. The companies in the best position are those that have more than one drug in development and a development pipeline in place. Right now the vast majority of drug candidates that go into human testing fail in the very early phases.

In addition, the worldwide market for pharmaceutical products, which includes biotechnology, is overwhelmingly driven by what happens in the United States. Not only is the United States by far the largest market, its growth rate is also the highest (see Figure 4). Significant problems would result if there were a societal referendum in the United States no longer allowing companies to be rewarded for tremendous innovation in developing drugs that minimize pain and suffering. I believe that we will continue to reward innovation, but it is important to keep in mind just how dominant the U.S. marketplace is in the overall economics of medicine.

A company has to make money if it is to survive; otherwise the result is just an endless consumption of cash. Over the past 10 years there has been a very significant acceleration in the number of profitable companies, which coincides with the long development time lines. We are seeing company after company turn profitable, and the list continues to grow.

How has the marketplace responded? As the number of profitable companies has accelerated, so has the amount of investment capital that has poured into the biotechnology sector. I believe the trend will continue, as there are many companies in this sector that will produce products to minimize or even eliminate the pain and suffering associated with human illness. Although a successful company that brings a breakthrough product to the marketplace can reap substantial rewards, it is important to remember that the sector will remain volatile.

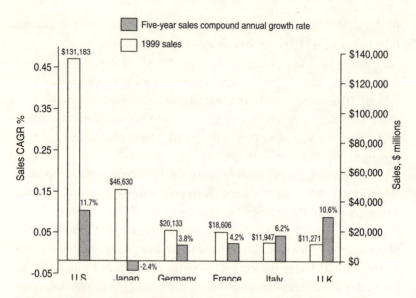

FIGURE 4 Five-year sales growth vs. 1999 sales by country.

Immunex, for example, introduced a breakthrough product for the treatment of rheumatoid arthritis, for which there have not been any new therapies in the past 10 years. Despite declines, had you invested in Immunex in 1997, you would have received a six- to seven-times return on your money. IDEC Pharmaceuticals is another example. IDEC introduced a breakthrough therapy for the treatment of non-Hodgkin's lymphoma. Although an investment in this company would have been volatile, a long-term steady investment would have resulted in a profit.

Over the next 5 to 10 years we will see tremendous growth in the pharmaceuticals industry because of substantial innovation and the secular trends mentioned earlier. A number of companies will succeed, some of which are just coming forward now with innovative therapies. This is an extremely volatile sector, but several of these companies will hit huge home runs. Although volatility presents danger, those who invest in and remain committed to the industry will earn enormous rewards.

Rebecca Eisenberg

The Role of Patents in Exploiting the Genome

The sequencing of the human genome is a great scientific accomplishment that opens the door to further scientific inquiry of a sort that would otherwise be impossible. In addition to being passionately interested in the patent issues this research presents, as a legal scholar I have a long-standing interest in the role of intellectual property in interactions between the public and private sectors and between universities and private firms in research science, with a focus on biomedical research. However, although the Human Genome Project has provided a rich terrain for exploring these issues, I am puzzled that intellectual property issues have become as prominent as they now are in public discourse regarding the genome project, particularly because patenting DNA sequences has been occurring for years and is certainly not a new practice.

DNA patenting began with little fanfare and controversy, in contrast to other expansions of the patent system, which have been extremely controversial during the same period. For example, a great deal of public controversy has occurred over the allowance of patents on microorganisms, animals, computer software, and, most recently, business methods. The issuance of patents in each of these areas promptly provoked opposition along with media commentary and congressional hearings. And in recent years, similar attention has been focused on the process of patenting genes, even though this did not occur when the first patent applications on genes were filed in the early days of the biotechnology industry about 20 years ago.

Thus, the practice of patenting genes was well established before it provoked any significant public controversy, which means that precedents had been set before the practice became questioned. The first public outcry over patenting DNA came in the early 1990s when the National Institutes of Health (NIH) filed patent applications on the first few thousand expressed sequence tags (ESTs, or gene fragments) to come out of the laboratory of Craig Venter when he was still at NIH. This provoked a great deal of controversy in the scientific community.

It also received a great deal of attention elsewhere, including antibiotechnology groups, and concern was signaled within the pharmaceutical industry from interests that generally favored the science but were uncomfortable with the patenting itself. More recently, stories about patenting genes have become almost a routine feature in media coverage of the Human Genome Project. Some stories are devoted entirely to intellectual property issues, while others focus on the so-called race between private- and public-sector initiatives to complete sequencing of the human genome.

Renewed Attention to DNA Patenting

So what has changed? For one thing the patent system generally is receiving more attention in public discourse than it did in the past, partly as a result of the boom in high technology and increased focus on the role of technology in the economy. Many questions have been raised about whether patents are hurting or helping progress in certain areas, particularly in information technology. In addition, our society is experiencing a period of profound ambivalence about the role of the private sector in matters relating to human health, something that was most conspicuous in the rhetoric about drug prices and pharmaceutical profits during the last presidential election.

The changing character of discovery in genetics and genomics also accounts for the shifting interest in patents, particularly regarding DNA sequences. In the early days of genomics patenting, gene patenting seemed to be a variation on patenting drugs, while it now appears to be more like patenting scientific information. There is also a clear history regarding why it makes sense to issue patents on drugs, and although some might contest this history, it does provide a clear case for patent protection. It is significantly less clear whether it makes sense to issue patents on scientific information.

Early Claims

The first generation of DNA sequence patents was directed toward particular genes that encoded certain proteins of interest, and we could identify these genes before anyone set out to clone them. For example, we knew a great deal about the insulin gene before cloning was attempted. Thus, the patent applications that were filed typically claimed the isolated and purified DNA sequence encoding the protein of interest, a recombinant vector that includes the DNA sequence, and a transformed host cell that includes the vector. All of these claims were framed in a way that distinguished them from naturally occurring products and covered tangible materials that were used to make therapeutic proteins, which were basically like other pharmaceutical products. A patent on the recombinant DNA starting materials would give a company an effective commercial monopoly on the recombinant proteins encoded by the DNA sequences.

In other words, having a patent on DNA sequences was similar to having one on a drug, although the gene patent was directed to the starting materials used in the production of the drug rather than to the drug itself. The Patent and Trademark Office, the agency that issues patents, and the courts treated patents on DNA sequences the same way they treated patents on new chemical compounds, or new drugs, looking to past cases that involved claims to new chemical compounds or newly isolated chemical compounds. The products-of-nature issue that seems to trouble many about gene patenting thus had been resolved before patenting occurred in cases involving isolated products, such as those that had been isolated from plasma or nature and that were made available in a form that served some human purpose. For example, there are old cases involving the patenting of aspirin, purified adrenaline, and vitamin B-12, all of which occur in nature. Therefore, the courts had no problem allowing the patenting of isolated and purified compositions that became available to meet a human purpose.

Motivations to Patent

The analogy to chemicals may never have been a perfect one, but it worked in the sense that it provided commercially effective patent protection that motivated investments in the development of new products. The biopharmaceutical industry is an area in which the patent system is important because it makes a difference in whether firms will invest in research. This is not the case in every industry.

Our patent system is a unitary one that purports to apply the same sets of rules across all fields of technology. But, in fact, those rules work very differently across different fields. In some industries, if you ask business managers or decisionmakers about the importance of patents to their motivation to invest in research and development, they will respond that it is not very important at all and that other factors are more significant in determining the profitability of their investments and innovations, such as being first to market or overcoming barriers to entry. In other words, the patents are just trading currency to get other patent holders to leave you alone. This is not so in the pharmaceutical industry, where there is empirical evidence to show that patents really do matter. This is because, according to the pharmaceutical industry, it costs a fortune to develop new drugs, with many costly failures for each successful product. Moreover, large regulatory costs are imposed in bringing new products to market. If competitors could enter the market for successful products and drive down their prices without having to incur development costs, including the costs of all the failures (the "free rider effect"), it would drive existing drug companies out of business.

The early biotechnology firms saw themselves as smarter higher technology pharmaceutical firms focused on developing therapeutic protein products. And they too wanted patents that would prevent free riders from destroying their profits. Patents on genes promised to provide protection from competition from free riders and allowed these new firms to raise capital and seek collaborators in the pharmaceutical industry. Some biotechnology firms still follow this essential business model, looking to identify and bring to market new therapeutic proteins either on their own or with their partners in the pharmaceutical industry. But the biotechnology and genomics industries have become much more diverse in their research and business strategies.

Some of the DNA sequences that emerge from the sequencing of the human genome will undoubtedly encode therapeutic proteins, such as insulin, and some firms will focus on identifying those proteins and bringing them to market. However, the primary value of the genome will not be the encoded instructions for producing therapeutic proteins. Rather, the genome will be a source of information for future research, some of which will ultimately lead to the development of products that are far removed from the genomic information that helped researchers along the path to drug discovery. And it is not obvious how patents on genes or on other bits of DNA sequence information can be used to capture the value that genomic information contributes to these other discoveries. Different participants in the biopharmaceutical research effort have very different perspectives that lead to different outlooks on the role of patents.

The pharmaceutical industry generally supports the patent system, but it is concerned about some of the genomics patents, and in recent years the pharmaceutical industry has invested in research to place genomic information in the public domain before the genomics firms can patent it. For example, the SNP Consortium, a group of major pharmaceutical firms, has been investing in identifying points of variance in the human genome and placing that information in the public domain. The Merck genome initiative is an effort by a private pharmaceutical firm to sponsor a university-based effort to create a catalog of fragments of genes and make that information freely available in the public domain. These private initiatives have provided an important reality check on the impact of the patent system, which motivates investment by allowing patent holders to charge monopoly prices. But, ultimately, it is the disaggregated consumers of end products who are paying those monopoly prices.

Trade-offs

The argument for patents in this situation is that without them consumers would not be able to benefit from a new product, and sometimes, but not always, this is true. In any event, consumers are not in a position to dispute this claim, although the prospect of Medicare drug coverage threatens to aggregate the interests of some of these consumers by consolidating them into a single powerful payer that would significantly alter the market for drugs.

Another way to view patents is that when they are issued for discoveries that are made on the road to drug development, they feed into future discoveries, or upstream innovations. The payoff that these patents promise to their owners will come from the pockets of future innovators. Most genomic discoveries are upstream inventions as opposed to downstream product developments, and they feed into a course of cumulative innovation. The trade-off presented by offering patent protection for these inventions is not simply how to balance the interests of consumers in low prices against the interests in creating incentives for further innovations but how to balance the interests of prior innovators against the interests of subsequent innovators. Another way to put it is that both the buyers and sellers of these upstream innovations are involved in the process of biomedical innovation.

Thus, the trade-off is between creating incentives and promoting and rewarding early-stage innovation versus creating incentives and motivating end-product development. From the perspective of the end-product developers, those who hold patents on these research inputs look like so many tax collectors, diluting their profits on potential new products. End-product developers are well organized politically, and they have a clear business model that includes a grounded view about the role of patents. The earlier upstream innovators are organized to some extent, but they are very diverse and are less clear about their business models and the role of patents in those models. They use their patents to raise investment capital in order to conduct research, and they hope that some of those patents will someday help them make a profit, perhaps by capturing a share of the profits made by subsequent innovators in developing new drugs.

The lack of clarity in the nexus between patents and potential profits is a problem, which may explain the overreaction in the financial markets that occurred when President Clinton and Britain's Prime Minister Blair made a rather tame announcement that they approved of the policy of placing genomic information in the public domain.

Who Decides?

If patents on genes are good for some innovators but bad for others, how do we know whether, on balance, they are promoting progress? In some ways this is always a guess. The patent laws reflect certain presumptions that offer some guidance, but these presumptions plainly are not true across the range of innovations that the patent system covers. In addition, the law is often indeterminate, and the Patent and Trademark Office and the courts must make some choices within a system that usually resolves such issues very slowly.

In genomics, for example, the Patent and Trademark Office is currently working through patent applications on ESTs, or gene fragments. These are relatively old discoveries, many of which were filed in the early 1990s. In determining what course to take, the Patent and Trademark Office looks to even older decisions based on older technologies for guidance. In many cases, however, the resolution of these issues is ultimately a policy decision. Although decisions can be appealed, this is a lengthy process that can take many more years. In addition, Patent and Trademark Office decisions are subject to review by the court of appeals for the federal circuit, and, more remotely, decisions of the federal circuit are subject to review in the U.S. Supreme Court. More remotely still, Congress can intervene at any time and change the rules.

Changing Views of What Can Be Patented

Although the patent statute includes a number of doctrinal levers for determining what can be patented and that constrain how the patent system responds to new technologies, genomics challenges some of the traditional tools for sorting through patent claims. Because we are 20 years into the biotechnology revolution, the landscape of discovery has shifted, and old cases offer limited guidance today.

A fundamental issue is one of how to patent DNA sequences. The statute says that a new process, machine, manufacture, or composition of matter can be patented. To date, DNA sequences have been patented as composition of matter, a characterization that emphasizes their material existence as tangible molecules. In this new high-throughput sequencing era, however, much of the value of newly identified sequences resides in the contribution they make to databases of information compared to their value as tangible molecules. Today, DNA sequences look more like information than molecules, as they are the tangible storage medium of cells, which use the information stored in their DNA to survive and reproduce. Newly sequenced DNA is stored in computer-readable form, and much of its value lies in making that information available to scientists.

Some patent applications are now pending that claim DNA sequences stored in machine-readable form. It is not clear what will happen to these patent applications and whether patents can be used to protect data. A few years ago the answer would clearly have been *no*. Now, it is not so clear.

The patent system has been expanding in many different directions as the courts try to accommodate information technology. With computers it is difficult to distinguish between machines and information, and it is also difficult to distinguish between compositions of matter and information in genomics. Allowing patents on information represents a major shift for the patent system, one that is probably unwise because patent rights are not well adapted to protecting information, particularly information about the natural world where independent discovery is inevitable. It is also not clear that patents are needed to motivate investment in genomics. Overprotection of information during the early stages may slow subsequent research more than it promotes original data collection.

Our old model of patents on genes as tantamount to patents on drugs worked well for the first generation of recombinant DNA products. Now, however, with genes looking more like information and providing an information base for drug discovery, a different business model is emerging, and it is not clear what role the patent system will play.

Troy Duster

Social Side Effects of the New Human Molecular Genetic Diagnostics

I take it as my task to talk about some of the social spin-offs of the mapping and sequencing of the human genome. And lest you think I regard the spin-offs as minor or trivial, let me use an analogy or at least an example.

The U.S. Department of Defense developed the Internet on behalf of communications around military and defense issues. No one could have predicted that in the short space of a few decades there would be something called a World Wide Web, which is a spin-off of the Defense Department's interest in the Internet. In other words, spin-offs can take center stage and move the original agenda to the side. I am not suggesting that that is going to happen with spin-offs of the Human Genome Project, but I want to alert us all to the fact that there are some implications of these spin-offs that are going to profoundly affect our lives in much the same ways that we now think about the dot-com revolution in Silicon Valley. I am going to draw a parallel connection between the computer revolution and what is happening in human genetics.

Those who justified federal funding of the Human Genome Project articulated a rationale that was related to improvements in health. Yet the general response to gene disorders in the population is somewhat peculiar. For the past six years I have been involved in a research project that examines how people in families with genetic disorders deal with their condition. This study involves over 300 people in various families where there has been a diagnosis of cystic fibrosis (a problem with the lungs and digestive system), sickle cell anemia (a blood disorder), or one of the thalassemias (also blood disorders). People in these families not surprisingly tend to know and care a lot more about scientific developments, such as the Human Genome Project, than the general population. Unless or until there has been a diagnosis, however, people tend to care very little about genetic disorders. It is a binary world—either you have it or you don't; you are involved in this very deeply or you are in the periphery. You may think this is true for all diseases. If it were, people who do not have heart disease, for example, would not care that much about cardiovascular disorders. But that is not true: people do watch their diets, lower their cholesterol counts, and exercise. So genetic disorders have a peculiar, more binary, frame to them.

What then really animates the general population about the Human Genome Project? The answer concerns whether genes place one in a certain category or perhaps explain some of our attributes or just possibly our behavior. In part, the link between molecular genetics and the popular imagination comes through an interesting route. People who live together in social and cultural groups for decades or centuries develop laws about who they can and cannot mate with and who they can and cannot marry. Endogamy laws are laws of culture and anthropology, not genetics. Who you can or cannot marry produces gene pools that collect over the centuries, which is why women who are ethnically Jewish have a greater risk for certain kinds of diseases than do others, or why people from the

TABLE 1   Ethnicities/ Groups Primarily Affected by Selected Disorder (USA)

Genetic Disorder	Groups Primarily Affected
Alpha-thalassemia	Chinese, Southeast Asian
Beta-thalassemia	Mediterranean
Tay-Sachs	Ashkenazi Jewish
Cystic fibrosis	Northern European
Phenylketonuria	European/Irish
Sickle-cell anemia	African American
Adult lactose deficiency	Chinese, African American
Duchenne muscular dystrophy	Northeastern British
Cleft lip/palate	Native American, Japanese

Source: L. Burhansstipanov, S. Giarratano, K. Koser, and J. Montgoven, 1987, *Prevention of Genetic and Birth Disorders*, California State Department of Education, Sacramento.

Mediterranean are at higher risk for beta-thalassemia, or why cystic fibrosis primarily affects those of Northern European ancestry, and why phenylketonuria primarily affects those of European descent (Irish in particular; see Table 1). Laws of endogamy and, of course, geography have determined the flow of genetic information and the prevalence of disease in certain populations.

Genetic tests have different sensitivities depending on the population being tested (see Table 2). The gene for cystic fibrosis was "discovered" about 10 years ago, but very quickly we found that there were over 250 mutations. The current test focuses on a common mutation, called ΔF508, but the test varies remarkably in its sensitivity in different groups. For example, among Caucasians there is a high rate of sensitivity; the test correctly assesses this particular allele over 9 times in 10. But for Asian Americans, in which cystic fibrosis has a much lower incidence, the test has much less sensitivity. At the very bottom are the Zuni Indians, who have a particular allele that predisposes them to cystic fibrosis, but that allele is not included in standard tests. In fact, there is no genetic test for the Zuni's version of cystic fibrosis. I propose to you that no test will be forthcoming because of the social location of the Zuni in our society.

Although the rationale for the Human Genome Project is health

TABLE 2   Genetic Epidemiology and Genotype/Phenotype Correlations for Cystic Fibrosis

Group	Incidence	Carrier Frequency	% Δ F508	Sensitivity
Caucasians	1:3,300	1:29	70	90
Ashkenazi Jews		1:29	30	97
Hispanics	1:8,500	1:46	46	57
African Americans	1:15,300	1:63	48	75
Asian Americans	1:32,100	1:90	30	30
Zuni	1:1,580			

and its delivery, when we come closer to the actual empirical cases what we are going to find—and we will see more of this in the next decade—is that biotechnology companies will develop certain kinds of gene tests and interventions with respect to drug therapies where there is a market. The money lies not with the Zuni but with the ΔF508 mutation in persons of Northern European descent.

Authenticity, Ethnicity, and Race

Molecular biologists and others in the biological sciences have been telling us for the past 20 years that there is no such thing as race, at least in any sense that is biologically meaningful. That is, we cannot find any kinds of physiological pathways, any kinds of circulation of the blood system, that correlate with what we typically see as the large racialized groups, such as Caucasians: nothing maps biologically.

From there we leap to the notion that race is therefore not a useful concept in biology. On the other hand, biological scientists are providing DNA profiles of persons likely to be in certain groups. Although at the level of the DNA we are remarkably similar, single nucleotide polymorphisms, or SNPs, can be important and significant. One of those particular kinds of SNPs might, for example, code for the protein for the clotting of the blood. A misspelling in the DNA can cause hemophilia. Therefore, it is important to look for particular kinds of polymorphisms to determine their function. Until about five years ago, this notion of finding an individual's SNP profile was so unlikely a procedure that it was not part of the landscape, but computer technology has made it a reality.

We can now put all of these SNPs on a computer chip—SNPs on chips. What the computer will permit us to do that we could not do before is to rotate those SNPs on chips and construct a DNA profile about differences and patterns and categories of persons. I take off my hat to the health possibilities promised by this capability. For example, we could group people with the same physical manifestation—for example, certain kinds of heart diseases—and look at their SNPs, identify the patterns, and arrive at a theory for their disease.

But there are other uses of this technology, and that is where I want to draw your attention. A few years ago some British forensic researchers, led by Ian Evett[1] published research in which they claimed that, by combining four to seven points along the DNA, certain kinds of allele frequencies could predict with 85 percent accuracy whether a person was either from the Caribbean islands or the United Kingdom. To scientists, that does not constitute race but rather an 85 percent chance that a person is either from the Caribbean or the United Kingdom. However, it is a proxy for race at the home office in Scotland Yard.

As we find more health applications of this technology in the coming years, the use of DNA typing for forensic uses and for determining countries and populations of origin will become increasingly familiar. We have already seen some of this in Illinois, where DNA profiling has been used to free people who were on Death Row.

An American group of researchers, doing a similar analysis, have corroborated Evett's work, suggesting that by combining particular allele frequencies one can predict patterns of population groups of which one is a member. Again, in a technical sense, this is not race; it is population groups and the likelihood of coming from a group or not. However, if you are a forensic scientist or a prosecuting attorney, these analyses provide a powerful weapon.

This use of SNPs on chips will head down another path, which is authenticity. The famous case of Thomas Jefferson and Sally Hemings provides an example. For years historians debated the link between Jefferson and Hemings. Just a couple of years ago, DNA analysis confirmed that Jefferson and Hemings produced children together, the descendants of whom are alive today. Historians said: "Well, if the DNA says so, it must be true that Jefferson really was the father of Hemings's offspring." Authenticity is coming at us through the mapping and sequencing. Whether one is authentically Native American has consequences in claiming fishing and land rights.

The same is true in Australia, where there is a big market in aboriginal art. It is also true in Canada with debates over First Nations people. In these nations, policies are focused on determining who is authentically Native American. In the past five years we have seen the initiation of research projects using DNA profiling to authenticate who is really Native American. This is not just fanciful. The Vermont legislature recently introduced a bill that asked for voluntary DNA profiling to determine authentic membership in a Native American population. I am suggesting that as we move along the path of more and more profiling of groups, we are going to find more of these requests for ethnic authenticity. The issue here, of course, is not health or gene disorders.

Race and Use of DNA Databases

I want to turn now to the DNA databases of the Federal Bureau of Investigations (FBI) and draw a connection to this type of research. Several years ago a few states started collecting DNA samples of convicted sex offenders. The purpose was straightforward: if one could find the possibility of drawing the connection, then one could have an easier conviction. A few years later, other states, including Virginia, joined in by collecting the DNA of all convicted felons, not just sex offenders. Now it is becoming more routine for states to collect samples of those arrested. We now have a national DNA database at the FBI.

Not long ago, the New York City Police Department, with the support of the mayor, pushed for a new type of crime control. It would equip police cars with little boxes no larger than a portable

CD player. If an officer stopped someone, he or she could take a saliva swab to obtain a DNA sample, take it back to the police car, put it on this device, and match it up to the database via satellite. Within 10 minutes the police would know if they had the right person. Because this is a standard procedure for pursuing people with outstanding tickets, why not do it with "outstanding DNA"? It is controversial, and then-Attorney General Janet Reno quickly moved to appoint a blue ribbon task force to study the problem.

I know from colleagues who are on that task force that they have concluded that this kind of police work could pass constitutional muster. Indeed, the New York City Police Department was allowed to engage in preliminary use of these tests. It is important to put this approach into the larger context of studies of racial profiling.

In terms of the biological sciences, there is no such thing as race. But for the New York City Police Department and other police departments that have engaged in systematic stopping of particular persons whose phenotype represents a particular group, there is such a thing as race. Consider the possibility that if police are stopping and arresting and using saliva swabs in one particular racial group because of racial profiling, we will soon find a considerable bias in the database. Therefore, if you are stopped and your saliva swab is taken and you are from one particular population group, your chances are astronomically higher of being in that database. My understanding again is that these kinds of strategies are going to be ratified and that we are probably going to find some version of this in the next short period.

Interestingly enough the United Nations Educational, Scientific, and Cultural Organization's statement on race,[2] published in 1995, is an admirable statement with the best of intentions. It says that race is not really an important biological category. One can admire the purpose behind it. It implies, however, that one should not be doing research on the topic of race. But while these professional societies of scientists are saying that race is not a phenomenon worthy of investigation, we have Evett in England and Shriver and his associates[3] in this country publishing research in scientific journals about allele frequencies in different ethnic population groups that serve as a practical proxy for race.

When I was a member of the Ethical, Legal, and Social Issues Advisory Committee of the federal Human Genome Project, some of us argued rather vociferously that issues of behavioral genetics deserve as much attention as those dealing with matters of health and medicine. Our concerns were borne out when an article was published in the *Journal of Neurology, Neurosurgery and Psychiatry* entitled "Crime and Huntingdon's Disease: A Study of Registered Offenses Among Patients, Relatives and Controls," using DNA that was obtained for the purpose of Huntington's disease research in an inappropriate way. In short, the gap between the behavioral and the medical applications of DNA profiling is narrowing.

Enhancement

I sit on the Advisory Panel on Germline Genetic Intervention of the American Association for the Advancement of Science. One of our members heard a report on National Public Radio suggesting that DNA technologies might be used by athletes to enhance their abilities. They went so far as to suggest that current scandals about Olympic athletes using drugs to enhance performance would pale before these new technologies. The committee member was skeptical, and so she inquired about the technical feasibility of this report. She was told that there are new collections of blood transfer protocols aimed at treating heart disease by encouraging blood vessel growth called the bio-bypass. It works in heart muscles but could work in other muscles as well. Another member of the committee responded by asserting: "Certainly, one could theoretically develop an element that permits expression of a more athero-protein in response to an environmental trigger—a kind of endogenous blood doping controlled by ingestion of an unmetabolized sugar."

Thus, this new technology could clearly be used for athletic enhancement. We have already seen diagnostic and therapeutic agents, such as human growth hormone for extreme short stature, being promoted to parents who just want taller children. The health value of this hormone risks being minimized by a combination of forces around social categories.

Ten years from now we will surely hear accounts of improved health as a result of all of this genetic research. But my prediction is that we will hear more about the forensic applications of this technology and its differential effect on different population groups. Indeed, it is likely to be the nonmedical uses of the new human molecular genetics that will come to dominate the impact these technologies will have on our lives.

Notes

1. I. W. Evett., I. S. Buckleton, A. Raymond, and H. Roberts, 1993, "The Evidential Value of DNA Profiles," *Journal of the Forensic Science Society*, 33(4):243-244; I. W. Evett, P. D. Gill, J. K. Scranage, and B. S. Wier, 1996, "Establishing the Robustness of Short-Tandem-Repeat Statistics for Forensic Application," *American Journal of Human Genetics*, 58:398-407.

2. S. H. Katz, 1995, "Is Race a Legitimate Concept for Science?" *The AAPA Revised Statement on Race: A Brief Analysis and Commentary*, University of Pennsylvania, Philadelphia.

3. Shriver, Mark D., Michael W. Smith, Ji Lin, Amy Marcini, Jousha M. Akey, Ranjan Deka, and Robert E. Ferrell, "Ethnic Affiliation Estimation by Use of Population-Specific DNA Markers," *American Journal of Human Genetics*, 60: 957-964, 1997.

Arthur L. Caplan

Mapping Morality
The Rights and Wrongs of Genomics

Several cases illustrate the types of ethical issues that are raised by the application of genomic information to the prevention, diagnosis, and treatment of human disease. These cases show that genetic testing is slightly different than other areas of medicine because it tells us things about people other than the individual being tested; it tells us things about the future that we might not want to know or make use of; and it may lead to the stereotyping of groups. Genetic testing is ethically and morally unique because many people define themselves according to their genes—that is, their race, their family, or the group to which they belong. When terms such as "blood" or "kinship" are used, you are talking in terms of genetics, and the more you learn, the more your sense of who you are may be jeopardized.

Genetic Testing Can Reveal More Than You May Want to Know

About nine months ago a man came to the University of Pennsylvania to be tested for the presence of the gene that is implicated in Huntington's disease. Although Huntington's disease has been around for a long time, it was not until about 10 years ago that a reliable genetic test became available for families in which this neurological disease exists. Children of affected individuals have a 50 percent chance of inheriting Huntington's, which does not strike until midlife or later. This man's father had succumbed to the disease following a terrible course.

After this individual was tested at our neurology institute, I received a telephone call and was told that the test had been done and that the results were good—he had not inherited the gene. I usually don't get these "good news" calls, so I was immediately suspicious. After asking why I was contacted, the caller said he needed my advice: not only had the test revealed that the man was not at risk, it had also shown that he was not biologically related to his father. We never told him that his father was not his biological father. Instead, we told him he was not at risk and sent him home. We changed the informed consent form the next day, however, to say that genetic tests could reveal facts about paternity.

Testing for Future Risks

In making a decision regarding whether to undergo such testing, people need to consider the implications of test results for insurance. Some choose to pay for such tests out of pocket so that their insurance company does not find out the results, because once such information is entered on the medical record, they might face problems obtaining insurance or maybe even employment, if they were flagged as being at risk for a disease that might be costly to treat or that might cause them to leave the work force early. However, not reporting the risk of disease to an insurance company or employer is fraud. If an insurer were to find out, the insured could find his or her policy voided.

In another case a man came to our clinic asking to have his daughter tested for increased risk of breast cancer. What made this case unique is that the daughter was only 11 years old. Breast cancer was prominent in the family; the man's mother and cousins had died of it, and he was concerned that he had passed on a predisposition to the disease to his daughter. As a precautionary measure he wanted to find out if she was at increased risk and, if so, have her breast buds removed. The issue in this case is that no one has the right to test that child involuntarily. She must be old enough to give her consent and decide for herself whether a prophylactic mastectomy is her choice. We did not test the girl.

Stigmatizing Groups

In a third case I was made aware of an advertisement recruiting members of the Jewish population of Baltimore, Maryland, for studies of depression. But because there is no evidence of more correlation in Jews than in any other group for depression, I called the researcher and asked him why he was targeting that group. His response was straightforward and simple. He found that population particularly compliant as research subjects and therefore good candidates for a study that would require several appointments.

Genetic Tests of Ancestry

There are two companies now advertising to do genetic studies to determine an individual's African ancestry, both of which are first attempting to recruit among African Americans. The goal is to offer people the option of determining what part of Africa their ancestors came from based on research that is fine tuning genetic analysis of different groups in Africa. Several questions immediately come to mind. Do people define themselves in a racial, ethnic, or cultural group by their genes? Or is it by their language, their customs, the way they speak, the way they dress, or the neighborhood in which they grew up? For example, many who are not Native American now want to be a Lumbee or a Chippewa so they can share in casino profits. A legislator in a Midwestern state came up with the idea that maybe people who claim to be members of a particular tribe should be genetically tested. The question then becomes: Could you be on a tribal council, or be an elder of the tribe, or be someone who grew up on a reservation and find out that you have too many "white genes" to be considered Chippewa, even though that is where you come from and you in fact have a leadership position in the tribe?

Genetics is fascinating because it tells us that there is a lot more commonality than difference among human beings and more commonality in nature generally. Simplistic census tabulations don't tell us how interconnected and interrelated we really are. Genetic information can be used to classify and lump, split and separate, identify and admit. Many nations have, for example, granted the right of return if you can show that your ancestors come from a particular place. Citizenship often keys on biological inheritance. In the future, genetics will intersect those social, scientific, anthropological, and even archeological areas in very interesting ways.

Where Do We Go From Here?

There are many other areas besides genetic testing that raise profound ethical issues, such as gene therapy, eugenics, germ-line therapy, and the targeting of drugs to people with particular biological types. Genetic testing, however, is here now. Are we ready to master it? One important step would be to ensure that people provide informed consent in advance of testing (which is not assured at this time) and that they fully realize the consequences of genetic testing in terms of insurance and other types of potential discrimination.

In addition, the privacy of genetic information must be assured. At this time it is not, which can result in employment or insurance disqualification if a person is found to have withheld information about preexisting risks for disease. Without blanket protections for privacy and confidentiality, people will be loath to undergo genetic testing that might be useful to them.

Policymakers are woefully ignorant about how to proceed in this area. Sometimes they do not even understand what DNA is or what genes are. And we are not ready to decide policy issues that control the production of genetically modified foods, or that determine how to prevent insurance discrimination on the basis of genetic profiles, or what should be done to ensure that people provide informed consent before genetic testing. To prevent the misuse of genetic information, we all need to be better informed about what it means and what it can tell us about ourselves. Otherwise, our values will be left far behind the technology.

Kathi E. Hanna

Summing Up
Finding Our Way Through the Revolution

> It has not escaped our notice that the specific pairing
> we have postulated immediately suggests a possible
> copying mechanism for the genetic material.
>
> —J. D. Watson and F. H. Crick

 This wonderful understatement printed in the journal *Nature* in 1953 changed biology forever. A combination of great intellect and luck led James Watson and Francis Crick to discover the molecular structure of DNA in the early 1950s. In retrospect, that discovery—that DNA was arranged in a double helical structure as a long chain of only four units called nucleotides—is startling in its simplicity. Yet 50 years later scientists are still trying to decipher the meaning of the complex code embodied in the straightforward patterns of these four units. The Human Genome Project, a grand-scale public and private effort to map and sequence the entire human genome, has been officially under way since 1990. Ten years later, on June 26, 2000, leaders of the federally and privately funded programs were invited to the White House to announce the

completion of a first rough draft of the human genome. The two initiatives had assembled 3.12 billion letters of the human genetic code using sequencers, sophisticated computer systems, and algorithms.

The White House event serves as a bright reminder that science often has to overcome skepticism, even from scientists, to move forward. When a massive human genome project was first contemplated 15 years ago, critics said the idea was absurd, impossible, even dangerous. In this volume, Leroy Hood describes how the persistence of scientists like Robert Sinsheimer, a biologist who was then chancellor of the University of California, Santa Cruz, envisioned this first "big science project" in biology. Not everyone shared his vision; it ran counter to the way biological research had been conducted for decades, which was in the form of small projects that were investigator initiated and not necessarily goals oriented. But people like Sinsheimer and Hood believed it was time for a different approach to biology, called "discovery science," in which scientists set out to see what they can see, with the goal of making sense of it later.

Harold Varmus, Eric Green, Leroy Hood, J. Craig Venter, and William Haseltine describe the evolution of this discovery process and how engineering and computer science have been essential in sorting out and making sense of the massive amounts of data contained in the genome. Eric Green describes how from 1990 to 1997 a central activity of the Human Genome Project involved constructing physical maps of the "24-volume human encyclopedia set." A major emphasis from 1998 to 1999 was the construction of page-by-page maps of each human chromosome, so-called second-generation physical maps of the human genome. With such rapid generation of the human genome sequences, the challenge has become learning how to assimilate all of these new data. All agree that the next key phase of the Human Genome Project is going to be the interpretation phase, analyzing the new sequence data and trying to figure out what those data mean. Eventually the Human Genome Project is expected to produce a sequence of DNA representing the functional blueprint and evolutionary history of the human species.

The Unity of All Living Things

In the early stages of the project, many scientists argued that biologists could never truly understand the human genome if they did not compare it to the genomes of other organisms. Much of Craig Venter's genomic work has focused on completing the genomic maps of lower organisms. He writes: "We all think that evolution involves adding on genetic information and complexity, but what we are finding is that most human pathogens probably started from a much more complex organism and eliminated genetic material during evolution." A fortunate finding for genome scientists has been that evolution has been remarkably conservative, retaining the same genes over and over again in different organisms. Nature's efficiency is manifest in the DNA of Earth's organisms. The genome project has provided a test of Darwin's theory of survival of the fittest. Certainly the fittest genes have survived across species and time, unless human intervention has altered or eliminated them. The power of genomic analysis is its ability to test the antiquity of our relationships with all other living things.

Leroy Hood refers to "evolution as a tinker," writing "there are numerous solutions to survival, but once a successful solution is reached, everything that happens subsequently is built on that successful solution." Thus, as genomic science moves forward, the study of other life forms can tell us in remarkable and surprising ways how we relate to other species and each other.

Arnold Levine and William Haseltine provide vivid examples of how the study of DNA reveals the unity of life on the planet. The fact that yeast contains the same genes as humans is such a simple evolutionary plan, yet so elegant. And although we tend to think of how lower organisms can be used to our own end, or to our destruction, consider the possibility that the insertion of a human gene into yeast could save its life. Haseltine uses the example of insulin—a human protein made by a gene—to illustrate how the unity of life can have practical consequences for humans. Modern technology provides the tools for slicing the insulin gene out of one particular individual and implanting it in a separate organism. Because of the unity of all living things, scientists can place that human gene into something that is removed from us by 2 billion years of evolutionary time—a bacterium or a yeast—and that organism will make the insulin for us. Barbara Schaal notes that by understanding the genome of cassava, or yuca, and other essential plant nutrients, we can find ways to help these species thrive to feed the world.

Evolution, Genomic Variation, and Social Consequences

Despite the remarkable similarities among the genomes of the earth's organisms, genomes can differ by variations in nucleotide sequences but also through duplications or deletions of DNA, through combinations that rearrange the order of genes, or by insertions of DNA that may have come from viruses. The process of sexual reproduction generates new combinations of genes—across multiple generations constituting the process of evolution.

With the human genetic code we find roughly 2 million to 3 million variations in the chromosomes. Millions of well-characterized so-called SNPs (single nucleotide polymorphisms) are now being used by scientists around the world to study linkage to disease. Mary Jeanne Kreek provides insight into how the study of these polymorphisms in the endorphin system of the brain and central nervous system could lead to prevention and treatments of diseases of these systems. She reports how studies of variants in opioids and opioid receptors might explain why people differ in their immediate response to pain or painful stimulus; immune, gastrointestinal, and cardiovascular and pulmonary function; and even mood, affect, cognition, and possibly learning and memory. In addition, study of these variations might lead us to why some individuals are at increased risk for myocardial infarction, certain forms of cancer, osteoporosis, and other chronic diseases. The pharmaceutical industry is extremely interested in using the study of SNPs to find ways to improve clinical trials and drug effects. This could lead to "personalized medicine," or pharmacogenetics.

But Troy Duster warns that a focus on what makes us different could also be misused to discriminate against vulnerable groups with shared characteristics. As biological scientists are able to provide DNA profiles of persons likely to be in certain groups, he cautions, the use of DNA typing for forensics and for determining countries and populations of origin will become increasingly familiar, leading toward a new kind of racial profiling that could have dangerous consequences. Duster also worries that pharmacogenetics might be disproportionately beneficial to the majority for a given condition, leaving those with more unique, and therefore less profitable, genomes behind. Daniel Kevles counsels us to remember the past when planning the future, citing the unpleasant history of the eugenics movement of recent history.

Douglas Wallace describes how the study of genomic variations among peoples, the Human Genome Diversity Project, can tell us the story of human history and migration. As the scientist Max Delbruck once said: "Any living cell carries within it the experiences of a billion years of experimentation by its ancestors." The Human Genome Project has provided insight into DNA as a historical molecule, writes Wallace. "It brings to us the information that arose billions of years ago and has been methodically passed down from generation to generation. So we are, in fact, the inheritors and recipients of all of those interesting evolutionary experiments. The information in your genome carries with it all of your prehistory. With that simple concept we can use DNA as the historical molecule to read out the history book of man and woman." By investigating the prevalence of certain genes that are maternally or paternally inherited, Wallace and others have demonstrated the movement of ancient civilizations out of Africa and into Asia, Europe, and eventually the New World.

No Simple Choices:
Genetic Testing for Human Disease

In this volume Harold Varmus and others remind us that nearly 50 years after Watson and Crick's discovery, genomics has become an integral and essential element of biotechnology and molecular biology, forging new perceptions of how life works, and changing our concept of our world and our origins. Applications of genomics are leading to new approaches to diagnosing and treating disease. Kris Jenner provides proof of the potentially vast market for biotechnology, which is seeking new products and profits from the "golden age of biology."

Cancer researchers hope the complete human genome sequence will provide information that could lead to cures. Over the past 15 years or so, researchers have learned that cancers are usually caused by the accumulation of several gene mutations, some of which activate cancer-promoting oncogenes, whereas others inactivate tumor suppressor genes. Genomics is providing insight into what turns genes off and on in the normal and abnormal cell cycles. Arnold Levine writes about the evolution of cancer genetics, which has led to the introduction of what is hoped to be the first in a long line of cancer drugs rationally designed—that is, intended to destroy cancer cells, not healthy cells, and based on an accurate knowledge of the initiation and progression of cancer. Extensive family history taking, pedigree analysis, and gene hunting led to the elucidation of a set of mutations responsible for many inherited cases of breast and ovarian cancer. Studies of these mutations that predominate in breast cancer also divulge information about ancient changes in our DNA. And while these mutations tell us something, they do not tell us everything. The variability of these mutations and the differences in their meanings complicate prescriptions for who should be tested for genetic predispositions.

Arthur Caplan explains why genetic testing is often considered unique from an ethical perspective for several reasons. First, in some cases it can provide information not only about the person being tested but also about his or her family members, who may or may not wish to know or have others know their risk status. Moreover, many of the diseases tested for predict risks of developing disease many years in the future. The psychosocial dimensions of this predictive knowledge are different from those relating to tests for concurrent disease.

In the context of presymptomatic, predictive, prenatal, or preconceptual testing, complex issues and risks can arise that might require a more involved informed-consent process. For example, the potential risks of insurance discrimination might be greatest for those who are currently healthy but who want to know whether they are at increased risk for a disease in the future. In addition, testing related to reproduction raises complex moral, psychological, and deeply personal issues. Genetic testing of children and adolescents who are at high risk of future disease because of family history raises special concerns, especially testing of asymptomatic children for genes for late-onset disorders, when there is no medical benefit, or for carrier testing when the information is not immediately useful for the child's reproductive decisionmaking.

Genetic Enhancement Versus Cure

Genetic enhancement is the attempt to make individuals better than well, optimizing their capabilities by taking them from standard levels of performance to peak performance. David Rothman believes this raises some intriguing questions because rather than make a copy of an individual, which is what would happen with cloning, genetic enhancement may be able to improve that individual, which might be more appealing. He writes: "Although the distinction between cure and enhancement has a surface logic, it has surprisingly little meaning in establishing a biomedical research agenda, in dictating medical practice, or in formulating health policy. To the contrary, cure and enhancement merge into each other and actually feed off each other, with interventions that begin in an effort to cure often quickly becoming enhancements."

Rothman speculates that "there is good reason to think that in the next 10 to 20 years we will have developed genetic enhancements to improve memory and perhaps problem-solving ability; to reduce dramatically the level of and need for sleep; to improve physical capacities to make us stronger and quicker; to provide perfect pitch; to provide personality traits, including higher levels of aggression or perhaps higher levels of altruism; to improve immunology and protections against diseases, such as cardiac disease and cancer; and to provide protections against weight gain and for increased longevity." He warns that these developments will not be without risk; therefore, society must demand that a calculus be developed to measure such risks and ensure a just distribution of the benefits.

Whose DNA Is It Anyway?

Our DNA holds secrets about our past and can help us predict the future. As such, it is the most powerful collection of information on Earth. The practice of patenting genes is well established, writes Rebecca Eisenberg, because it is well recognized that this information can be used to promote innovation and profit. However, she recognizes that because patents on some genes might be good for some innovators but bad for others, it is difficult to know whether patents, on balance, promote progress. Current patent laws reflect certain presumptions that offer some guidance about this balance, but these presumptions are not always true across the range of innovations that the patent system covers. In addition, the law is often indeterminate, and the Patent and Trademark Office and the courts must continue to make choices to ensure that the patent system promotes the benefits of genomics without curtailing or unjustly restricting progress.

Conclusion

This volume reflects the great promise of the Human Genome Project despite the possibility for its misuse. The knowledge gained could cure cancer, prevent heart disease, and feed millions. At the same time, its improper use can discriminate, stigmatize, and cheapen life through frivolous enhancement technologies. Because of the promise for great good, we all need to understand more about the science and application of human genomics to ensure that the harms do not materialize. An informed citizenry is essential to making the right choices. In his collection of essays, *A Passion for DNA* (Cold Spring Harbor, NY: Cold Spring Harbor Laboratory Press, 2000), James

Watson writes: "Moving forward will not be for the faint of heart. But if the next century witnesses failures let it be because our science is not yet up to the job, not because we don't have the courage to make less random the sometimes most unfair courses of human evolution."

Watson's optimism should always be followed with a healthy degree of skepticism—signs of good science and a thoughtful society. In the early days of the project, Watson was asked to describe its goals. In his typical blunt manner he answered, "to find out what being human is." The simplicity of that response, not unlike the simplicity of the double helix, underlies a complexity that we will continue to explore in the twenty-first century and beyond.

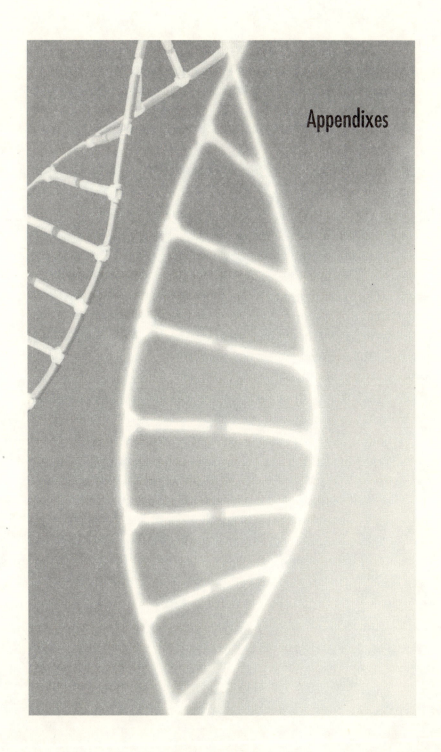

Appendixes

Appendix A:
Contributor
Biographical Sketches

ROBERT BAZELL
Chief Science Correspondent, NBC News

During his career with NBC News, Robert Bazell has reported on a wide range of subjects in science, technology, and medicine, from throughout the United States and around the world. He was awarded the prestigious George Foster Peabody Award for distinguished achievement and meritorious service in broadcasting. Mr. Bazell's extensive tracking of the AIDS epidemic, which began in 1982, has included reports from the United States, Africa, Europe, the Caribbean, and South America, and earned the Alfred I. duPont-Columbia Award and the Maggie Award from the Planned Parenthood Federation. Recently, Mr. Bazell won an Emmy in the Outstanding Informational or Cultural Programming category for his in-depth report on experimental brain surgery. Mr. Bazell received a B.A. in biochemistry from the University of California at Berkeley, did graduate work in biology at the University of Sussex, England, and was awarded a doctoral candidate degree in immunology at Berkeley. Most recently, he has written a book, *HER-2: The Making of Herceptin, a Revolutionary Treatment for Breast Cancer* (Random House, October 1998).

ARTHUR L. CAPLAN
Director, Center for Bioethics and Trustee Professor, University of Pennsylvania

Arthur L. Caplan has been the Director of the Center for Bioethics and Trustee Professor at the University of Pennsylvania since 1994. He is also Professor of Molecular and Cellular Engineering, Professor of Philosophy, and Chief of the Division of Bioethics at the University of Pennsylvania Medical Center. Prior to joining the University of Pennsylvania, Dr. Caplan served as Professor and Director of the Center for Biomedical Ethics at the University of Minnesota. He is currently Chairman of the Advisory Committee to the Department of Health and Human Services, Centers for Disease Control, and Food and Drug Administration on Blood Safety and Availability, and a member of the boards of Celera Genomics, Medscape, the National Center for Policy Research for Women and Families, and the National Disease Research Interchange. His recent books include *Ethics and Organ Transplants* (Prometheus, 1999), *Am I My Brother's Keeper?* (Indiana University Press, 1998), *Due Consideration: Controversy in an Age of Medical Miracles* (John Wiley & Sons, 1997), and *Prescribing Our Future: Ethical Challenges in Genetic Counseling* (Aldine Press, 1993). In addition, he is the author of more than 475 articles and reviews in professional journals, and lectures widely. Dr. Caplan was the first president of the American Association of Bioethics and is a fellow of the American Association for the Advancement of Science.

ROB DESALLE
Curator and Co-Director of the Molecular Laboratories, American Museum of Natural History

Rob DeSalle is a curator in the American Museum's Division of Invertebrate Zoology and co-director of the Molecular Laboratories. Dr. DeSalle's fields of specialization include molecular evolution, population genetics, molecular systematics, and developmental biology. His research utilizes molecular genetic approaches to study problems in evolution and the application of systematic techniques to genomics. The focus of his research has been on the molecular systematics of the *Drosophilidae*. In addition, Dr. DeSalle is one of the founders of the Museum's Conservation Genetics program, which applies studies at the molecular level to the conservation of wildlife and wild lands throughout the world. In 1996, Dr. DeSalle and his colleagues developed a genetic test for caviar that helped gain protection for sturgeon in the Caspian Sea basin under the Convention on the International Trade in Endangered Species (CITES). Dr. DeSalle has curated several exhibitions at the Museum including the highly praised "Epidemic!" and "The Genomic Revolution." He received his B.A. in biology from the University of Chicago and his Ph.D. in biological sciences from Washington University. In addition to his research and teaching, Dr. DeSalle co-authored *The Science in Jurassic Park* (Basic Books, 1997) and has published widely in scientific journals including *Nature* and *Science*. He is co-author of "Gene family evolution and homology: Genomics meets phylogenetics" in the *Annual Review of Genomics and Human Genetics* (Press, 2000).

TROY DUSTER
Professor of Sociology, New York University

Troy Duster is currently Professor of Sociology at New York University. He is also a member of the American Association for the Advancement of Science Committee on Germ-Line Intervention. From 1996 to 1998, Dr. Duster served as chair of the joint National Institutes of Health-Department of Energy (NIH/DOE) advisory committee on Ethical, Legal, and Social Issues in the Human Genome Project (the ELSI Working Group). He served as Chairman of the Department of Sociology at the University of California, Berkeley, from 1986 to 1989, and is the University's former Director of the Institute for the Study of Social Change. Dr. Duster is a former member of the Assembly of Behavioral and Social Sciences of the National Academy of Sciences, and he has served on the Committee on Social and Ethical Impact of Advances in Biomedicine at the Institute of Medicine. He is the author of numerous books and monographs including *The Legislation of Morality* (Free Press, 1970), *Cultural Perspectives on Biological Knowledge* (co-edited with Karen Garrett, 1984), and *Backdoor to Eugenics* (Routledge, 1990). His works have also appeared in *Politics and the Life Sciences, The Genetic Frontier: Ethics, Law and Policy,* and *DNA and Crime: Applications of Molecular Biology in Forensics*. His most recent publication on this topic is "The Social Consequences of Genetic Disclosure," in Ronald Carson and Mark Rothstein, eds., *Culture and Biology* (Johns Hopkins University Press, 1999).

REBECCA S. EISENBERG
Robert and Barbara Luciano Professor of Law, University of Michigan Law School

Rebecca S. Eisenberg joined the University of Michigan Law School faculty in 1984 and has taught courses in intellectual property, torts, the legal regulation of science, and legal issues in the Human Genome Project. Professor Eisenberg has written extensively about patent law as applied to biotechnology and the role of intellectual property at the public-private divide in research science, publishing in scientific journals and law reviews. She has received grants from the Program on Ethical, Legal, and Social Implications of the Human Genome Project from the U.S. Department of Energy Office of Biological and Environmental Research for her work on private appropriation and public dissemination of DNA sequence information. Professor Eisenberg has also played an active role in public policy debates concerning the role of intellectual property in biomedical research. In 1996, she chaired a workshop on intellectual property rights and research tools in molecular biology at the National Academy of Sciences, and in 1997 to 1998, she chaired a working group on research tools for the National Institutes of Health. She is a member of the Advisory Committee to the Director of the National Institutes of Health and a past member of the Working Group on Ethical, Legal, and Social Implications of Human Genome Research. She is a graduate of Stanford University and Boalt Hall School of Law at the University of California, Berkeley.

ELLEN V. FUTTER
President, American Museum of Natural History

Ellen V. Futter has been president of the American Museum of Natural History since November 1993. She previously served for 13 years as President of Barnard College, where, at the time of her inauguration, she was the youngest person to assume the presidency of a major American college. She is director of a number of organizations and has a strong record of public service, including having served as chairman of the Federal Reserve Bank of New York and on the boards of The Legal Aid Society and the American Association of Higher Education. Ms. Futter is a fellow of the American Academy of Arts and Sciences, a member of the Council on Foreign Relations, a partner and a member of the Executive Committee of the New York City Partnership, Inc., and a member of the Executive Committee of NYC & Company. She is widely recognized as a dynamic voice for education and is an active supporter of women's issues. She has been awarded numerous honorary degrees and is the recipient of the National Institute of Social Science's Gold Medal Award. Ms. Futter was graduated Phi Beta Kappa, *magna cum laude*, from Barnard College in 1971. She earned her J.D. degree from Columbia University's Law School in 1974.

ERIC GREEN
Chief, Genome Technology Branch, National Human Genome Research Institute, Director, NIH Intramural Sequencing Center

Eric Green received his M.D. and Ph.D. from Washington University School of Medicine (St. Louis, Missouri) in 1987. During his residency training in clinical pathology, he worked in the laboratory of Maynard Olson, where he developed approaches for utilizing yeast artificial chromosome (YACs) to construct physical maps of DNA. His work also included initiation of a project to construct a complete physical map of human chromosome 7 within the Washington University Genome Center—one of the first centers funded as part of the Human Genome Project. In 1992, he became an assistant professor of pathology, genetics, and medicine as well as a co-investigator in the Human Genome Center at Washington University. In 1994, he moved his research laboratory to the intramural program of the National Human Genome Research Institute at the National Institutes of Health (Bethesda, Maryland), where he now serves as Head of the Physical Mapping Section, Chief of the Genome Technology Branch, and Director of the National Institutes of Health Intramural Sequencing Center (NISC). Dr. Green's research focuses on the mapping and sequencing of mammalian genomes and the isolation and characterization of genes causing genetic diseases.

KATHI E. HANNA

Kathi E. Hanna is a science and health policy consultant, writer, and editor specializing in biomedical research policy and bioethics. She has served as Research Director and Senior Editorial Consultant to the National Bioethics Advisory Commission and as Senior Advisor to the President's Advisory Committee on Gulf War Veterans Illnesses. In the 1980s and early 1990s Dr. Hanna was a Senior Analyst at the now defunct congressional Office of Technology Assessment, contributing to numerous science policy studies requested by committees of the House and Senate on science education, research funding, biotechnology, women's health, human genetics, bioethics, and reproductive technologies. In the past decade, she has served as an analyst and editorial consultant to the Howard Hughes Medical Institute, the National Institutes of Health, the Institute of Medicine, and several charitable foundations. Before coming to Washington, she was the Genetics Coordinator at Children's Memorial Hospital in Chicago, where she directed clinical counseling and coordinated an international research program investigating prenatal diagnosis of cystic fibrosis. Dr. Hanna received her A.B. in Biology from Lafayette College, M.S. in Human Genetics from Sarah Lawrence College, and a Ph.D. from the School of Business and Public Management, George Washington University.

WILLIAM A. HASELTINE
Chairman and CEO, Human Genome Sciences, Inc.

William Haseltine is Chairman of the Board of Directors and Chief Executive Officer of Human Genome Sciences, Inc., a company he founded in 1992. Human Genome Sciences' mission is to develop products to prevent, treat, and cure disease, based on its leadership in the discovery and understanding of human genes. Dr. Haseltine holds a doctorate from Harvard University in biophysics. He was a professor at Dana-Farber Cancer Institute, Harvard Medical School and Harvard School of Public Health from 1976 to 1993 before joining Human Genome Sciences. Dr. Haseltine has a distinguished record of achievement in cancer and AIDS research and has received numerous honors and awards for his achievements in science, medicine, and business. He is the founder and Editor of the online journal *E-Biomed: The Journal of Regenerative Medicine*, and was formerly the Editor-in-Chief of the *Journal of AIDS*. He is on the editorial boards of many other scientific journals. Dr. Haseltine has over 250 publications in the scientific literature and has been awarded more than 50 patents for his discoveries. He is a 1996 recipient of the American Academy of Achievement Golden Plate Award.

LEROY HOOD
President and Director, Institute for Systems Biology

Leroy Hood is recognized as one of the world's leading scientists in molecular biotechnology and genomics. His professional career began at California Institute of Technology, where he and his colleagues pioneered four instruments that constitute the technological foundation for contemporary molecular biology. One of the instruments has revolutionized genomics by allowing the rapid automated sequencing of DNA. Since then, Dr. Hood's research has focused on the study of molecular immunology and biotechnology. He moved to the University of Washington in 1992 to create the cross-disciplinary Department of Molecular Biology, applying his laboratory's expertise in DNA mapping to the analysis of human and mouse immune receptors as well as initiating investigations in other areas. In 1999, he founded the Institute for Systems Biology to pioneer systems approaches to biology and medicine. Dr. Hood was one of the first advocates for and is a key player in the Human Genome Project. He also played a leading role in deciphering the secrets of antibody diversity. Dr. Hood holds numerous patents and awards for his scientific breakthroughs and prides himself on his life-long commitment to making science accessible to the general public, especially children. Dr. Hood is a member of the National Academy of Sciences, the American Philosophical Society, and the American Academy of Arts and Sciences. He has published more than 500 peer-reviewed papers and co-authored numerous textbooks and other works.

KRIS H. JENNER
Portfolio Manager, T. Rowe Price Associates, Inc.

Kris H. Jenner is Vice President, Portfolio Manager, and Investment Analyst with T. Rowe Price Associates, Inc., specializing in the coverage of biotechnology and pharmaceutical companies. He is Chairman of the Investment Advisory Committee and a vice president of the Health Sciences Fund. Dr. Jenner is also a Vice President and Investment Advisory Committee member for the Blue Chip Growth Fund, Growth Stock Fund, Mid-Cap Growth Fund, and New Horizons Fund. Prior to joining the firm in 1997, Dr. Jenner worked at the Laboratory of Biological Cancer, The Brigham & Women's Hospital, and Harvard Medical School. He earned his B.S. in chemistry from the University of Illinois, a D.Phil from Oxford University, England, and an M.D. from the Johns Hopkins School of Medicine.

DANIEL J. KEVLES
Stanley Woodward Professor of History, Yale University

Daniel J. Kevles has written extensively about issues in science and society, past and present. He is the author of several prizewinning books, including most recently *The Baltimore Case: A Trial of Politics, Science, and Character* (W.W. Norton & Company, 1998), and is the co-editor with Leroy Hood of *The Code of Codes: Scientific and Social Issues in the Human Genome Project* (Harvard University Press, 1992). His articles, essays, and reviews have appeared in scholarly and popular journals, including the *New Yorker,* the *New York Review of Books,* the *New York Times*, and the *Los Angeles Times*. Dr. Kevles received his B.A. in physics and Ph.D. in history from Princeton University. From 1964 to 2001, Dr. Kevles taught history at the California Insitute of Technology, where he was Koepfli Professor of the Humanities and directed the Program in Science, Ethics, and Public Policy. In 2001 he was appointed Stanley Woodward Professor of History at Yale University.

MARY JEANNE KREEK
Professor and Head, Laboratory of the Biology of Addictive Diseases, The Rockefeller University

Mary Jeanne Kreek is Professor and Head of the Laboratory of the Biology of Addictive Diseases at The Rockefeller University and Senior Physician of the Rockefeller University Hospital in New York City. She is also Principal Investigator and Scientific Director of an National Institutes of Health-NIDA Research Center, "Treatment of

Addictions: Biological Correlates," which has been ongoing since 1987. After joining the Rockefeller Institute in 1964, Dr. Kreek, along with Dr. Vincent P. Dole and the late Dr. Marie Nyswander, performed the initial studies of the use of a long-acting opioid agonist, methadone, in chronic management of heroin addiction—studies which led to the development of the first effective pharmacotherapy for treatment of an addiction. Currently, the Laboratory of the Biology of Addictive Diseases includes a multidisciplinary team of molecular biologists, analytical chemists, neuroscientists, physicians, including psychiatrists, internists, and others working to study the molecular and behavioral neurobiology of addictive diseases and related clinical neurobiology of addictions. Since 1994, Dr. Kreek's work has been expanded to include the study of human and molecular genetics, including studies of polymorphisms of genes of special interest and family studies for which she was recently awarded an NIH-NIDA human molecular genetics grant along with the continuation of another collaborative grant focused on the mu opioid receptor gene. She is author of over 300 scientific reports. She received an honorary doctorate degree from the University of Uppsala in 2000.

ARNOLD J. LEVINE
President, The Rockefeller University

Arnold J. Levine, a leading authority on the molecular basis of cancer, became President of The Rockefeller University in December 1998 and was named the University's first Robert and Harriet Heilbrun Professor of Cancer Biology. His research focuses on the tumor suppressor gene called p53 and on its protein product, which he discovered in 1979. Now studied in laboratories worldwide, p53 is helping to develop a new generation of cancer therapies. Dr. Levine continues his research as head of Rockefeller's Laboratory of Cancer Biology. He came to Rockefeller University from Princeton University, where he was the Harry C. Wiess Professor of Life Sciences. Between 1984 and 1996, Dr. Levine presided over a major expansion of Princeton's life sciences program as Chairman of the Department of Molecular Biology. He helped shape U.S. science priorities as Chairman of an influential 1996 review panel on federal AIDS research funding. Dr. Levine was elected to membership in the National Academy of Sciences in 1991 and its Institute of Medicine in 1995. Among his numerous awards are the Katharine Berkan Judd Award from Memorial Sloan-Kettering Cancer Center, the 1994 Bristol-Meyers Squibb Award for Distinguished Achievement in Cancer Research, and the First Annual Strang Award from the Strang Cancer Prevention Center, also in 1994. He is currently a member of the scientific and medical advisory boards of the Howard Hughes Medical Institute, a trustee of Cold Spring Harbor Laboratory, and serves on the executive committee at the University of Pennsylvania. Dr. Levine is the author of the book *Viruses* (Scientific American Library, 1992).

DAVID J. ROTHMAN
Bernard Schoenberg Professor of Social Medicine and Director of the Center for the Study of Society and Medicine, College of Physicians & Surgeons of Columbia University

David J. Rothman is a social and medical historian. His first work, *The Discovery of the Asylum* (Little Brown, 1971), traced the early history of caretaker and custodial institutions and won the American Historical Association's Albert Beveridge Prize. In 1983, Dr. Rothman joined the Columbia medical school faculty, and his current research has explored the history of health care institutions as well as health policy and practices. His recent work includes *Strangers at the Bedside* (Basic Books, 1991), analyzing how law and bioethics transformed medical decision making, and *Beginnings Count* (Oxford University Press, 1997), which traces the technological imperative in health care. Dr. Rothman's current articles examine the relevance of medical professionalism and the ethics of human experimentation. Under a grant from the Human Genome Project, he and Sheila M. Rothman have analyzed the social implications of genetic enhancement technologies. Their forthcoming book is titled *Remaking the Self* (Pantheon, 2002). Dr. Rothman has had a particular interest in ethics, human rights, and medicine and has written extensively on the ethics of human experimentation. He chairs the Open Society Institute Program on Medicine as a Profession.

SHEILA M. ROTHMAN
Professor of Public Health in the Division of Sociomedical Sciences at Columbia's Mailman School of Public Health
Deputy Director of the Center for the Study of Society and Medicine at the Columbia College of Physicians and Surgeons

Sheila M. Rothman's current research focuses on genetics, with a special interest in understanding the impact of new genetic knowledge on group identity. She is coprincipal investigator with David J. Rothman on "The Genome Project and the Technologies of Enhancement" (National Institutes of Health). Its goal is to identify and analyze the challenges that genetic enhancements pose for American health policy and social policy. She presently serves as a member of the Task Force on Human Genetics at the Columbia College of Physicians and Surgeons and was recently appointed chairman of the Task Force on Genetics and Public Health at the Mailman School of Public Health. Her books include *Woman's Proper Place: A History of Changing Ideals and Practices, 1870 to the Present* (Basic Books, 1978) and *The Willowbrook Wars: A Decade Long Struggle for Social Justice,* coauthored with David Rothman (HarperCollins, 1984). Her most recent book, *Living in the Shadow of Death: Tuberculosis and the Social Experience of Illness in American History* (Basic Books, 1994), analyzes the lives of people with tuberculosis, and traces the impact of disease and public health policies on individuals, their physicians, and the larger community. Sheila Rothman is also interested in the links between human rights and medicine. Together with David J. Rothman, she has published articles on how AIDS came to Romania and medical accountability in Zimbabwe in the *New York Review of Books*. Since 1995, she has been a member of the Bellagio Task Force on Securing Bodily Integrity for the Socially Disadvantaged: Strategies for Controlling the Traffic in Organs for Transplantation. She is the principal investigator of a multisite study that is analyzing patient and family decisionmaking toward organ donation in the United States.

BARBARA A. SCHAAL
Professor of Biology, Washington University

Barbara A. Schaal's research is in the area of plant evolutionary genetics. Her current projects include studies of plant domestication, the genetics of invasive plants, and the genetic architecture of disease resistance in plants. From 1993 to 1997, Dr. Schaal served as Chairman of the biology department at Washington University, where she is currently Spencer T. Olin Professor of Biology and Professor of Genetics in the medical school. She has been Executive Vice President of the Society for the Study of Evolution and President of the Botanical Society of America. She has served on numerous journal editorial boards and is an associate editor of *Molecular Biology and Evolution, Conservation Genetics*, and the *American Journal of Botany*. She has been a Guggenheim fellow and is an elected fellow of the American Association for the Advancement of Science and the National Academy of Sciences. She currently co-chairs the National Research Council Standing Committee on Biotechnology, Agriculture and the Environment. Dr. Schaal received her undergraduate degree at the University of Illinois at Chicago and her Ph.D. from Yale University.

HAROLD VARMUS
President, Memorial Sloan-Kettering Cancer Center

Harold Varmus, the former Director of the National Institutes of Health and co-recipient of a Nobel Prize for studies of the genetic basis of cancer, is currently the President and Chief Executive Officer of Memorial Sloan-Kettering Cancer Center. Much of Dr. Varmus's scientific work was conducted during his 23 years as a faculty member at the University of California, San Francisco, where he,

Dr. J. Michael Bishop, and their co-workers demonstrated the cellular origins of the oncogene of a chicken retrovirus. This discovery led to the isolation of many cellular genes that normally control growth and development and are frequently mutated in human cancer. For this work, Dr. Bishop and Dr. Varmus received many awards, including the 1989 Nobel Prize for Physiology or Medicine. In 1993, Dr. Varmus was named by President Clinton to serve as the Director of the National Institutes of Health (NIH), a position he held until the end of 1999. During his tenure at the NIH, he initiated many changes in the conduct of intramural and extramural research programs and helped increase the NIH budget from under $11 billion to nearly $18 billion. In addition to authoring over 300 scientific papers and four books, Dr. Varmus currently serves on the World Health Organization's Commission on Macroeconomics and Health and a National Research Council panel on genetically modified organisms. He has been a member of the National Academy of Sciences since 1984 and of the Institute of Medicine since 1991.

J. CRAIG VENTER
President, The Center for the Advancement of Genomics, the Institute for Biological Energy Alternatives, and the J. Craig Venter Science Foundation

J. Craig Venter, Ph.D., is highly regarded for his major scientific contributions to genomic research. His three newly formed not-for-profit organizations are dedicated to exploring social and ethical issues in genomics, and to seeking alternative energy solutions through microbial sources. Dr. Venter began his formal education after a tour of duty in Vietnam from 1967 to 1968. After earning a bachelor's degree in biochemistry and a Ph.D. in physiology and pharmacology, both from the University of California at San Diego, he became a professor at the State University of New York at Buffalo and the Roswell Park Cancer Institute. He then moved to the National Institutes of Health campus where he developed expressed sequence tags, or ESTs, a revolutionary new strategy for gene discovery. In 1992, he founded The Institute for Genomic Research (TIGR). There he and his team decoded the genome of the first free-living organism, the bacterium *Haemophilus influenzae,* using his new whole genome shotgun technique. In 1998 Dr. Venter founded Celera Genomics to sequence the human genome using the techniques developed at TIGR along with new mathematical algorithms and automated DNA sequencing machines. The successful completion of this research culminated with the publication of the human genome in February 2001 in the journal *Science.* Dr. Venter is the author of more than 220 research articles. He is the recipient of numerous honorary degrees and scientific awards including the 2001 Paul Ehrlich and Ludwig Darmstaedter Prize. Dr. Venter was recently elected to membership in the National Academy of Sciences and is also a member of the American Academy of Arts and Sciences and the American Society for Microbiology.

NICHOLAS WADE
Science Editor, *The New York Times*

Nicholas Wade has been a science reporter at the *New York Times* since 1998. Prior to that, he served on the *Times*'s editorial board and as science editor. Formerly a reporter with *Science* magazine, Mr. Wade was also Washington correspondent and deputy editor of *Nature.* He is the author of several books including *Life Script,* an account of the human genome (Simon & Schuster, 2001). He has contributed articles to the *Wall Street Journal,* the *Washington Post,* the *Times of London* and the *New Republic.* Mr. Wade received a B.A. in natural sciences in 1964 from King's College in Cambridge, England.

MICHAEL WALDHOLZ
Deputy Editor, Health and Science, *The Wall Street Journal*

In June 1980, Michael Waldholz joined the *Wall Street Journal* as a reporter in New York covering medicine and healthcare and pharmaceutical industries. He was named a senior special writer in March 1994, became a news editor for the science, technology, and health group in May 1995, and was named Deputy Editor for health and science in January 1996. In 1997, Mr. Waldholz led a team of *Wall Street Journal* reporters that was awarded a Pulitzer Prize in the national-reporting category for chronicling the development and effects of new AIDS therapies. Also in 1997, he and *Wall Street Journal* reporter David Sanford won the National Association of Science Writers' Science-in-Society Journalism Award in the newspaper category for their series of articles focusing on new AIDS therapies. Mr. Waldholz is the author of *Curing Cancer* (Simon & Schuster, 1997) and is a co-author of *Genome*, a book about the hunt for human genes (Simon & Schuster, 1990). Mr. Waldholz appears weekly on CNBC's "Power Lunch," reporting about health and biotechnology.

DOUGLAS C. WALLACE
Donald Bren Professor of Molecular Medicine
Director, Center for Molecular and Mitochondrial
Medicine and Genetics, Colleges of Medicine and
Biological Sciences, University of California, Irvine

Dr. Douglas C. Wallace was born in Maryland in 1946. After growing up in Maryland and New York, he completed his Bachelor of Science degree at Cornell University in Ithaca, New York. Following two years in the service, he moved to Yale University where he completed Master's and Doctorate of Philosophy degrees by 1975. After one year of postdoctoral study at Yale, he joined the faculty at Stanford University as Assistant Professor of Genetics and remained in that position until 1983. He then moved to Emory University in Atlanta, Georgia as Professor of Biochemistry and Associate Professor of Neurology, Pediatrics, and Anthropology. In 1990 he was appointed the Robert W. Woodruff Professor of Molecular Genetics and Director of the Center for Molecular Medicine. He was also founding Chairman of the Department of Genetics and Molecular Medicine from 1992 to 1995. In 2002, Dr. Wallace moved to the

University of California, Irvine as the Donald Bren Professor of Molecular Medicine to found a new Center for Molecular and Mitochondrial Medicine and Genetics. Throughout his career, Dr. Wallace and his team have studied the genetics of human and mammalian mitochondrial genes encoded in either the mitochondrial DNA (mtDNA) or nuclear DNA (nDNA), elucidated the origins and ancient migrations of our species, and demonstrated the role of mitochondrial gene variation in a variety of degenerative diseases, aging, and cancer. In 1986, Dr. Wallace's work on mtDNA variation and human origins was featured in the NOVA program "Daughters of Eve" and was reported in a 1988 cover article of *Newsweek*, "The Search for Adam and Eve." In 1994 he was awarded the William Allan Award by the American Society of Human Genetics and in 1995 was elected to the National Academy of Sciences.

MICHAEL YUDELL

Michael Yudell, MPH, M.Phil., received his masters in public health from Columbia University's Mailman School of Public Health. He has been a Health Policy Analyst at the National Institute of Environmental Health Sciences and is now a member of the molecular laboratories at the American Museum of Natural History. Yudell is currently working on his second book, *Welcome to the Genome: A User's Guide to the Genetic Future,* which he is coauthoring with Rob DeSalle. This book is to be published next year by John Wiley & Sons. In 2003 he is expected to complete his doctorate in the program of History and Ethics of Public Health and Medicine at Columbia University.

Appendix B:
Conference Schedule

Friday, September 22, IMAX™ Theater, American Museum of Natural History

9:00 a.m. Welcome and Introduction
ELLEN V. FUTTER, *President, American Museum of Natural History*
DR. MACK LIPKIN DISTINGUISHED LECTURE
What Does Knowing about Genomes Mean for Science and Society?
HAROLD VARMUS, *President, Memorial Sloan-Kettering Cancer Center*

Session 1:   **New Frontiers in Science, Technology, and Law**

10:00   Introduction
Nicholas Wade, *Science Editor,* The New York Times

10:15   Sequencing the Human Genome: Elucidating Our
Genetic Blueprint
Eric Green, *Chief, Genome Technology Branch,*
National Human Genome Research Institute
Director, NIH Intramural Sequencing Center

10:45   After the Genome: Where Should We Go?
Leroy Hood, *President and Director, Institute for*
Systems Biology

11:15   Break

11:30   The Role of Patents in Exploiting the Genome
Rebecca S. Eisenberg, *Robert and Barbara Luciano*
Professor of Law, University of Michigan Law School

12:00 p.m.   Questions from the Audience

12:30-2:00   Lunch Break

Session 2:   **New Perspectives on Genetic Disease**

2:00   Introduction
Robert Bazell, *Chief Science Correspondent,*
NBC News

2:15   The Origins of Cancer and the Human Genome
Arnold J. Levine, *President, The Rockefeller University*

2:45   Genetic Analysis of Breast and Ovarian Cancer
        Mary-Claire King, *American Cancer Society Research
        Professor of Genetics, Departments of Medicine and
        Genetics, University of Washington*

3:15   Break

3:30   Social Side-Effects of the New Human Molecular
        Genetic Diagnostics
        Troy Duster, *Professor of Sociology, New York University*

4:00   Questions from the Audience

4:30   Break

4:45   Plenary Address
        The Essential and Non-Essential Nature of the
        Human Genome
        Stephen Jay Gould, *Agassiz Professor of Zoology,
        Harvard University*
        *Vincent Astor Visiting Research Professor of Biology,
        New York University*
        *Frederick P. Rose Honorary Curator, American Museum of
        Natural History*

Saturday, September 23, IMAX™ Theater, American Museum of
Natural History

9:00 a.m.   Introduction
            Ellen V. Futter, *President, American Museum of
            Natural History*

            Opening Address
            James Watson, *President, Cold Spring Harbor
            Laboratory*

Session 3:   Exploring Human Variation: Understanding Identity
               in the Genomic Era

10:00   Introduction
               ROB DESALLE, *Curator and Co-Director of the*
               Molecular Laboratories, American Museum of Natural
               History

10:05   Using Maternal and Paternal Genes to Unlock
               Human History
               DOUGLAS C. WALLACE, *Robert W. Woodruff Professor of*
               Molecular Genetics
               Professor and Chairman, Department of Genetics and
               Molecular Medicine
               Director, Center for Molecular Medicine, Emory University
               School of Medicine

10:30   Eugenics, The Genome, and Human Rights
               DANIEL KEVLES, *Professor of Humanities, California*
               Institute of Technology
               Visiting Professor of History, Yale University

10:55   Break

11:10   Redesigning the Self: The Promise and Perils of
               Genetic Enhancement
               DAVID J. ROTHMAN, *Bernard Schoenberg Professor of*
               Social Medicine and Director of the Center for the Study
               of Society and Medicine, College of Physicians &
               Surgeons of Columbia University

11:35   Gene Diversity in the Endorphin System: SNPs,
               Chips and Possible Implications
               MARY JEANNE KREEK, *Professor and Head, Laboratory of*
               the Biology of Addictive Diseases, The Rockefeller
               University

12:00 p.m.   Questions from the Audience

12:30-2:00   Lunch Break

Session 4: Genomics and Biotechnology: Opportunities and Challenges in the 21st Century Marketplace

2:00   Introduction
            MICHAEL WALDHOLZ, *Deputy Editor, Health and Science,* The Wall Street Journal

2:10   Genomics: A Rapid Road from Gene to Patient
            WILLIAM HASELTINE, *Chairman and CEO, Human Genome Sciences*

2:40   Genomics, Biotechnology, and Agriculture
            BARBARA A. SCHAAL, *Professor of Biology, Washington University*

3:10   Break

3:20   Investing in the Biotechnology Sector
            KRIS H. JENNER, *Portfolio Manager, T. Rowe Price Associates, Inc.*

3:50   Mapping Morality: The Rights and Wrongs of Genomics
            ARTHUR L. CAPLAN, *Director, Center for Bioethics and Trustee Professor, University of Pennsylvania*

4:20   Questions from the Audience

4:50   CONCLUDING REMARKS
            MICHAEL J. NOVACEK, *Senior Vice-President and Provost of Science, American Museum of Natural History*

Index